DISTILLING KNOWLEDGE

NEW HISTORIES OF SCIENCE, TECHNOLOGY,
AND MEDICINE

SERIES EDITORS
Margaret C. Jacob, Spencer R. Weart, and Harold J. Cook

BRUCE T. MORAN

DISTILLING
KNOWLEDGE

ALCHEMY, CHEMISTRY, AND THE
SCIENTIFIC REVOLUTION

HARVARD UNIVERSITY PRESS

CAMBRIDGE, MASSACHUSETTS

LONDON, ENGLAND

First Harvard University Press paperback edition, 2006

Library of Congress Cataloging-in-Publication Data
Moran, Bruce T.
Distilling knowledge : alchemy, chemistry, and the scientific
revolution / Bruce T. Moran.
p. cm. — (New histories of science, technology, and medicine)
Includes bibliographical references and index.
ISBN-13 978-0-674-01495-4 (cloth)
ISBN-10 0-674-01495-2 (cloth)
ISBN-13 978-0-674-02249-2 (pbk.)
ISBN-10 0-674-02249-1 (pbk.)
1. Chemistry—History.   2. Alchemy—History.
3. Science, Renaissance.   I. Title.   II. Series.
QD15.M67 2004
540′.9—dc22   2004052601

FOR BARBARA, KATE, AND RASHMI

CONTENTS

DISTILLING KNOWLEDGE

There is something that does not quite make sense about including a subject like alchemy in a discussion of scientific revolution. Science, after all, is rational and ordered. Alchemy, we think, presumes disorder and irrationality. Common sense tells us that there are certain classes of objects, certain kinds of knowledge, and certain ways to go about discovering truths of nature that can be regarded as "scientific." Alchemy, because of its associations with magic and the occult, certainly does not belong here. To see things differently would be either crazy or intellectually counterfeit. Science, we know, is a particular form of knowledge made up of experimental facts, impartial observations, and specific theories. Anyone, no matter what his or her background, who would honestly seek objective truths in nature would ultimately have to reach the same conclusions about how the world works—right?

But here is where lines separating the rational and the absurd get a little fuzzy, and also where the well-defined intellectual image of science gets a bit scuffed up by rubbing against the texture of real life. The problem is that during the period of discovery and theoretical change called the Scientific Revolution of the sixteenth and seventeenth centuries, not everyone started out with the same assumptions when they attempted to represent what was going on in nature. Neither did everyone share the same sort of lived experience. What might have seemed clear and objective from one point of view could have seemed altogether unreasonable from another. ⌈Moreover, if the history of science means paying attention not only to the creation of a certain form of knowledge but also giving credit to various ways in which practical experience led to insights about the operation of nature, then a variety of activities, some of them learned and bookish, some of them requiring the skillful use of hands to pursue what now appear to be improbable goals, are relevant to its discussion.⌉Alchemy, although motivated by assumptions about nature not shared by many today, still occasioned an intense practical involvement with minerals, metals, and the making of medicines. Alchemical procedures produced effects and led to the analysis of various parts of the natural world. So, rather than cutting away the scientific lean from the presumed pseudoscientific fat when carving up natural knowledge in the "early modern" world, we should try to understand how both fat and lean worked together to support intellectual life and to promote the process of discovery. ⌈In that way we can begin to comprehend how diverse and even contradictory ways of explaining the operations of nature were sometimes intertwined as they sought to unravel nature's secrets. ⌉

Various perceptions of nature coexisted during the sixteenth and seventeenth centuries, and exploring how each made sense of natural phenomena and sought to explain relationships between objects of nature lets us develop a greater depth of field in picturing the era itself.⌈That contemplating the history of science should in some way be related to an examination of culture is not just the recommendation of an historian advocating ethnic sensitivity in evaluating how natural knowledge is created⌉Actually, someone else, a real scientist whose writings reflected a thoroughly practical frame of mind, a Harvard professor and medical doctor named William James, also knew that in any historical period claims to reason, experience, factual knowledge, and objectivity were inevitably related to the ways people preferred to see the world. Reason, he knew, was a great means to an end; but when it came to selecting a destination for the vehicle of reason, something else was driving the bus.

There he is, in surviving nineteenth-century photographs, with a massive and scraggly beard. James turned to philosophy after years of studying chemistry, comparative anatomy, and physiology. Though he was an empiricist and a pragmatist in both psychology and philosophy, he was nevertheless able to argue for the relevance, along with logic and rational insight, of subjective convictions, better known as the passions, in influencing and sometimes in determining intellectual choice and claims to certainty. No one knew better than James the risks involved in taking this view. The physical sciences, he recognized, were overwhelming in their utility and reliance on objective evidence. Allowing a place for sentimental preference or esoteric beliefs in making claims to knowledge was, he thought, on the one hand, "silly," and on the other, "vile." "When one turns to the magnificent edifice of the physical sciences," he wrote, "and sees how it was reared; what thousands of disinterested moral lives of men lie buried in its mere foundations . . . what submission to the icy laws of outer fact are wrought into its very stones and mortar; how absolutely impersonal it stands in its very augustness—then how besotted and contemptible seems every little sentimentalist who comes blowing his voluntary smokewreaths, and pretending to decide things from out of his private dream!" (James, 1896; rept. 1979: 17).

Well, that looks definite enough. Science makes truth and personal opinion is frivolous. And yet, James conceded that when wishful thinking and arcane wisdom were banished from science, what one was left with in the quest for certainty was still not pure reason. After all, both the magician-astrologer and the biochemist could, relative to their own networks of belief, claim to be objective in observing and interpreting the world. "The greatest empiricists among us," James observed, "are only empiricists upon reflection."

As far as objectivity was concerned, he was even more cautious. "When, indeed, one remembers that the most striking practical application to life of the doctrine of objective certitude has been the conscientious labors of the Holy Office of the Inquisition, one feels less tempted than ever to lend the doctrine a respectful ear" (pp. 21, 23). What one does for good reason and with a sense of objective certainty, in other words, follows from a willingness to believe in something. The same rule applies regardless of whether one's willingness to believe embraces doctrines of religion, the principles of alchemy, or the precepts of scientific method.

In thinking about the history of science, most of us are accustomed to believing in the authority of a "grand narrative," the story of the triumph of human reason over mysticism, magic, and the occult. The major battle in this exalted conflict, one in which the brotherhood of reason finally dispelled the orcs of intellectual darkness, took place, according to the story line, during the Scientific Revolution of the sixteenth and seventeenth centuries. After that, astrologers and alchemists awakened from their enchanted sleep and became astronomers and chemists. You don't have to be an expert in the history of science to suspect that there was more to the tale than this, however. As with the too-tidy telling of any tale, when things look neat and well ordered in retrospect, we may wonder if we are learning more about what we want to see than what is really there. What would happen if we could find a way to drop into the sixteenth and seventeenth centuries and see the world from the historical inside out? Would the grand narrative still ring true for us, or would the metaphor of Scientific Revolution need to be adjusted as the result of our experience? One thing would be certain. We would discover the prominence of alchemical theory and practice in debates about how nature works, and we would also become aware that those skilled in alchemical procedures were contributing to the creation of natural knowledge by sometimes getting their hands dirty and manipulating different substances in order to produce new effects. Quite possibly, we would also see that certain aspects of scientific reasoning were themselves being discovered as a result.

I have to admit that, as far as the relevance of alchemy to the history of science is concerned, mine is not a completely neutral point of view. I have what some would call a "pet hypothesis." It is not simply that I think that certain alchemical operations like distillation and sublimation influenced the work of later professional chemists, or that significant figures of the Scientific Revolution like Robert Boyle and Isaac Newton pursued alchemical projects. Both things are true, are well documented, and have a place in what follows. However, what I want to give attention to is something else—something a little less obvious, but every bit as important. What I want to do is to step outside the grand narrative of the victory of reason over nonsense and to consider the interdependence of supposed opposites in the creation of new learning during the sixteenth and seventeenth centuries.

But wait, you say. Alchemy makes a nice anecdote, but it is a fable—at best, a romantic fantasy. It may be beguiling, but it leads to nothing. Moreover, it has no useful purpose and, as a knowledge system, has no means to perpetuate itself didactically. How can this relate to science? We know that what makes science beautiful is method; and in this sense alchemy, which presumably has no method, gets ugly fast. My answer is simply this. Most of us have a very imprecise, if not an altogether cockeyed, view of what early modern alchemy was all about. In fact, most of what we think we know has been created for us by other generations with specific cultural axes to grind. In regard to the relation between alchemy and the history of science, earlier accounts were often concerned to make the history of science appear to be essentially modern history (another kind of grand narrative). The attributes allotted to alchemy were thus assigned to the superfluous part of a pair of opposites (reason versus superstition) in which preference for real power and utility could get acknowledged and presumed romance and rubbish rejected. But this way of thinking about alchemy only expressed a desire to establish counterpoint rather than continuity in the ways that the early modern era described and interacted with nature. Indeed, numerous historians have been at work in the last half century to show that alchemy was a rational subject, did have utilitarian value, did develop according to prescribed procedures, and could be taught, even if at times its language was obscure.

What one calls science embraces a very large area—so large, in fact, that it sometimes admits of paradox. Alchemy may seem an ironic element in discussions of scientific revolution; but the relationship between the two becomes particularly obvious when, along with the standard objects studied by historians of science such as motions, matter, personalities, theories, and discoveries, we begin to consider physical processes and practical experiences themselves, the doing and making of something through personal agency, as appropriate objects of discussion. Constructive procedures of many sorts (making, handling, and transforming things for purposes of curiosity and utility) claimed the attention of artisans and academics, men as well as women, in the early modern world and offered the means to express attitudes and values about nature in the act of causing things to happen. When viewed as part of the history of alchemy and chemistry, the practices of artisans can tell us a great deal about the variety of opinions concerning how nature operates and what the appropriate means of influencing nature might be. And here is the most important thing. Even when their procedures and projects lacked success, the involvement with alchemical and chemical processes by numerous figures across the social spectrum had implications for further knowledge because, unless altogether accidental, to do and to know what to do were, and still are, connected. That relationship, the connection between action and knowing, helps establish what has been called in a very different context a "region of transformation," an intellectual space that admits of new possibilities in which interpretations of experience, failed as well as successful attempts to make things, and even the impact of the emotions converge in the messy act of conceiving knowledge (Bollas, 1987: 28). Simply put, the history of science does not always have to be written as a giant success story. There can be room for the experience of *both* frustration *and* gratification—even when the subject is the Scientific Revolution. In fact, if we ignore how experiences are interwoven, lay as well as learned, satisfying as well as disappointing, we stand a good chance of missing what is really going on when accepted forms of knowledge begin to change. The scene is not one of light overcoming darkness, but of an animated muddle of belief, disillusion, and reinterpretation that is all part of negotiating what there is to talk about in the structure of nature, and how best to learn more about it. The subject of alchemy stands center stage here; and before going any further, we have to know more exactly what alchemy is. What does it mean in the early modern world?

CHAPTER ONE

DOING ALCHEMY

If you were young and inquisitive in 1597, because you were born into a world familiar with the earth-moving astronomical theories of Nicolaus Copernicus (1473–1543), you might have read the dazzling defense of Copernican astronomy just published by Johannes Kepler (1571–1630) called the *Cosmographic Mystery* (1596). You might also have known of a terrific book printed in that year that historians of a later time would revere as the first real textbook in the history of chemistry. The book, which was written by a German physician, poet, and teacher of high school–age boys named Andreas Libavius (ca. 1555–1616), promised to explain the composition and properties of bodies to the youth of the day, and to do so by means of exact descriptions of chemical procedures. Although hard to read in Latin, it would have taught you how to prepare a variety of chemical substances and how to make them purer by means of fire. You could have learned about assaying techniques, how to analyze minerals and metals, and how to make medicines out of them as well. The author made plain how chemical changes followed from combining different substances, explained quantitative methods for determining alloys, described the use of balances and weights, and gave precise instructions about how to build and use a variety of laboratory vessels. In a later edition, the book contained plans for constructing a chemical workshop and included a wealth of illustrations depicting all sorts of glassware, furnaces, and laboratory apparatus. In either version, however, you would have found a wonderfully pragmatic and logical guide to the useful, empirical, and theoretical parts of an art long discussed by scholars and daily practiced by experienced adepts. My guess is that you would have been excited by all of this and would not have been the least bit put off, disappointed, or confused by the book's title, namely, *Alchemy* (Libavius, 1597).

Practitioners of alchemy were among the most ardent investigators of nature before and during the period of the Scientific Revolution; and to understand the relationship of alchemy to the pursuit of natural knowledge, we first have to get a feel for the variety of projects in which they were involved and for the different schools of thought to which they adhered. Some were enthusiastic about making gold and silver; some focused more on making medicines. Still others sought out new procedures in developing a variety of chemical technologies. Some found room to do all these things at once. Some were physicians or philosophers who enjoyed the privileges of university degrees. Others were artisans who learned their art close to home. Some were itinerant and lived on the margins of society, while others enjoyed civic rights or held courtly appointments. Some were Moslems, and others were Jews or Christians. Some were women, others men; some were sincere, others frauds. Lots of people were involved with alchemy in late Renaissance and early modern Europe; and, with the exception of the deceptive and crackbrained among them, no one should think that what they were up to was either frivolous or uninformed.

A serious and practical pursuit is probably not what occurs to most people when they think of alchemy. That is because the subject has acquired, mostly due to the efforts of the solid citizens of modernity living on the Enlightenment side of town, a very shabby appearance. Like crystals that are shaped by their places in the earth, alchemy has been formed and twisted by the historical spaces in which it has been forced to live. The eighteenth and nineteenth centuries, for instance, developed a real sensitivity for intellectual boundaries, and in this respect those things identified as being spiritual and other things marked as physical all of a sudden needed to stay out of each other's neighborhoods. Given the enforced separation, alchemy was sent to live with its metaphysically batty great aunts. The part of the family tree linked to esotericism and mystical excess then sadly defined the whole activity; and the subject itself, earlier characterized by empirical expertise and utilitarian promise, fell into categories labeled occult, magic, or superstition—realms of belief, maybe, but certainly not divisions of science. The problem is that historically these categories really missed the point as far as alchemy was concerned, because alchemy was never altogether anything that people *believed* in; it was something that people *did*. And it is from the view of doing alchemy, that is, of actively responding to nature so as to make things happen without necessarily having the proven answer for why they happen, that a certain passion could on occasion combine quite well with empirical inquiry and practical desires so as to suggest new possibilities for natural knowledge.

This is what Libavius is *doing* in his *Alchemy*, and it is the reason why some historians have wished to raise a monument to him in the pantheon of great chemists. To do so, however, really misrepresents the world that Libavius lived in, because, while doing all the above, this "great chemist" was also busy defending the art of transmutation, deliberating on the contents of the Philosophers' Stone, and explaining the secret meanings of ancient hieroglyphs, enigmas, and symbols. The stones that built Libavius's alchemy, an alchemy that looks to some so modern, came from structures whose foundations were actually very old. Indeed, part of his book dealt with what were called "magisteries," substances whose external impurities had been removed so they could be used as powerful medicines. One of the best ways to prepare magisteries was by means of distillation; and although Libavius paid attention to lots of other chemical procedures to produce chemical extracts, when making his magisteries, he followed a long alchemical tradition in which the primary procedure was distillation and the principal purpose was to make the purest substance of all, something linked, it was thought, to the first stuff of creation, and sometimes given the name "the fifth essence."

✳ Galileo once referred to wine poetically as "light held together by moisture." He may not have known it, but he had actually expressed a very old alchemical opinion, one that acknowledged the existence in wine, indeed in all of nature, of something truly celestial, pure, and life-enhancing; and something that might be got at by means of distillation (in other words, through separating and condensing the more volatile parts of a mixture into liquids). In Europe, some of the most important figures in this tradition of distillation alchemy included a number of Franciscan friars. One came from France and was called John of Rupescissa (died ca. 1366). Another, named Raymond Lull (ca. 1234–1315), came from Catalonia, although none of the alchemical works bearing his name was actually written by him; and a third was a thirteenth-century Englishman who spent lots of time in Paris named Roger Bacon (1214?–1294). Both John and Roger apparently also spent lots of time in jail, or at least in some sort of imposed confinement, but that is another story. What they were all looking for was a super-medicine, an elixir or *aqua vitae* that could purify physical bodies of their impurities, rid the human body of disease, and prolong life. The means of finding this elixir was disputed, but one tradition was based on the work of an Arabic writer who tells us that his name is Jabir ibn Hayyan. In medieval Latin texts, Jabir is called Geber. Jabir thought that the best way to separate the parts of nature was by means of distillation. Bringing the distillates from a variety of substances together, he claimed, would yield the elixir itself. In Western Europe, alchemists of the fourteenth century began to think that the elixir was not so much a product of combining different distilled ingredients but was instead the end product of a series of distillations (usually of one substance only) gradually increasing in purity. The most pure substance of all (a universal medicine created by art and found nowhere in nature) had lots of names, but the one that was used most often was the fifth essence (Multhauf, 1954, 1956). One of the most important procedures for producing the fifth essence began with the distillation of wine.

There is something about doing distillation that combines action and reflection in such a way as to produce a feeling of unity and knowing. Primo Levi, author, chemist, and survivor of Auschwitz, wrote in his book *The Periodic Table* (1975): "Distilling is beautiful. First of all because it is a slow, philosophic, and silent occupation, which keeps you busy but gives you time to think of other things, somewhat like riding a bike. Then, because it involves a metamorphosis from liquid to vapor (invisible), and from this once again to liquid; but in this double journey, up and down, purity is obtained, an ambiguous and fascinating condition, which starts with chemistry and goes very far. And finally, when you set about distilling, you acquire the consciousness of repeating a ritual consecrated by the centuries, almost a religious act, in which from imperfect material you obtain the essence . . . and in the first place alcohol, which gladdens the spirit and warms the heart" (Levi, 1984: 57–58).

People had been distilling alcohol long before 1300, but it was around that time that alcohol began to appear in the alchemical literature under names like burning water, the water of life, and the fifth essence. An early reference to distilled alcohol appears in the work of a man named Salernus, a member of the faculty of the famous medical school at Salerno, around 1100. Peter of Spain discussed it at the close of his *Marvellous Treatise on Waters*, and the medical and anatomical writer Thaddeaus Alderotti showed how to redistill alcohol from wine by means of a coiled condensing tube that passed through a condensing trough. Indeed, there was always something truly remarkable about the properties of alcohol. It was, first of all, something of a physical contradiction: a water that burned. When placed in alcohol, animal and vegetable matter tended to receive extra life—at least it appeared not to rot or putrefy quite so quickly. Most important for alchemists whose main interest was in distillation, alcohol could dissolve materials such as resins and essential oils that were not dissolvable in water. This was a substance that seemed spiritual (look how fast it evaporates) and life enforcing, not to say invigorating; and, for some, it was by no means difficult to think of it as a less pure remnant of the vitalizing first heavenly matter of creation. In this regard, if one were to look for the source of a variety of useful medicines, fragrant oils, perfumes, and even strengthening liquors such as that prepared by the Benedictine monk Dom Bernardo Vincelli in 1510 and still available off the shelf today under the label of Benedictine Brandy, one would have to look to a technology that was altogether recognized as alchemical. The medieval theologian Albert the Great understood this perfectly. That is why he viewed distillation as one of the most important methods employed by alchemists. Three hundred years later, while the Scientific Revolution was well underway, that view had hardly changed. When, for instance, the sixteenth-century naturalist Conrad Gesner (1516–1565) wished to describe remedies of various sorts, he separated those derived through distillation from those that were "non-alchemical," that is, not distilled or sublimed (Forbes, 1948; Abrahams, 1971) (see Figure 1).

Most often it was the distillation of herbs that shaped the processes of medicinal alchemists, but medical fifth essences could also be derived from other materials. In this regard ancient sources furnished much information about a number of mineral distillates, including mineral acids in various grades of purity that were some-

Figure 1. From Christoph Wirsung, *Ein New Artzney Buch* (Newstadt an der Hardt, 1582) showing the preparation of medicines from plants using a "gallery" furnace for multiple distillations. University of Wisconsin Library.

times described as "waters," such as the highly corrosive "strong water" (*aqua fortis* or nitric acid), and sometimes as "oils," like oil of vitriol (in other words, sulphuric acid). In preparing these substances, alchemical practitioners depended on a variety of instruments such as alembics, cucurbits, retorts, and furnaces. Not many of these vessels remain intact today. Some exist as fragments, and others, like the ubiquitously pictured "pelican" (a vessel used for recirculating distillates), were probably more frequently described than actually constructed or employed (Anderson, 2000). Nevertheless, with their help the entire mineral kingdom could be added to the list of potential medicines, and the fifth essences of mercury, antimony, gold, and other inorganic materials extracted. It was easy to think that the world itself comprised a magnificent and abundant pharmacopoeia, and that the means by which the doors to this great pharmacy could be opened was to be found in the practical operations of distillation alchemy. This is certainly the meaning behind Hieronymus Brunschwig's (ca. 1440–ca. 1512) definition of distillation in his famous *Book Concerning the Art of Distilling* (Strassburg, 1500). There he notes that "distilling is nothing other than purifying the gross from the subtle and the subtle from the gross . . . with the intent that the corruptible shall be made incorruptible . . . and the subtle spirit be made more subtle so that it can better pierce and pass through the body . . . [and can be] . . . conveyed to the place [in the body] most needful of health and comfort" (Brunschwig, ca. 1530; rept. 1971: 9) (see Figure 2).

Brunschwig thus linked alchemy with the process of separating the pure, medicinal parts of a substance from parts that were considered harmful, poisonous, or impure. That notion became an important feature of later medical chemistry and, as we will see, enjoyed rebirth as a vital component of a particular doctrine of Renaissance medicine. In the Middle Ages it governed many of the alchemical descriptions of the threesome we have already met: Rupescissa, Lull, and Roger Bacon.

Figure 2. A pipe filled with cold water makes a fanciful cooling tower. Solutions are heated at the bottom and a distilled medicinal water (an *aqua vitae*) is collected at the top. The depiction of alchemical instruments sometimes arose more from the imagination than from actual use. From H. Brunschwig, *Das Buch zu Distillieren* (Strassburg, 1532). University of Wisconsin Library.

✳ Around the middle of the fourteenth century, John of Rupescissa was making predictions. He liked making predictions, or *prophetica*, and composed thirty of them, of which only five remain. Most of all he liked to predict the return of Christ and his personal rule on earth. In 1356 he predicted that the final days were at hand and would begin four years hence. For a number of years thereafter, one could expect widespread catastrophe—earthquakes, storms, famine, plague, even monsters running amok. Moreover, as if things were not bad enough, this would also be a time in which an anti-Christ from the West replaced one from the East. One would have to wait until 1367 for things to get better when an angelic pope brought the rule of the anti-Christ to an end and paved the way for the long-expected restoration of the world. Ideas like these placed John at variance with established religious authorities. But what superiors really disliked was his rigid reading and defense of Franciscan rules emphasizing poverty, and it was this that most likely led to his house arrest at the Cloister Figeac in the French countryside. More significant to us, however, is that, when not writing about the renovation of creation, John was composing alchemical texts—two of them, in fact, one called *A book Concerning the Contemplation of the Fifth Essence of All Things* and another, less certainly linked to him, called *The Book of Light* (Benzenhöfer, 1989: 12ff).

The first text, sometimes simply known as *Concerning the Fifth Essence*, was very popular and has been found to exist in over 130 manuscripts and in numerous printed editions from the fourteenth to the eighteenth centuries. The first Latin printed text appeared in 1561. The important thing about it, however, is not its longevity, but its assertion that the fifth essence extracted from terrestrial things was, when prepared correctly, very much like the stuff that comprised the heavenly spheres. Like others, John was convinced that the fifth essence was in fact a super-medicine and possessed heavenly powers that would assure the health of the body and prolong life. Extracting the fifth essence from things was really to extract "star stuff" and each metal, he reasoned, contained a heavenly essence that corresponded to a particular planet and acted on a particular part of the body.

For Rupescissa, the actual material from which the fifth essence was to be derived did not matter as much as the procedure or process applied in preparing it. In other words, the source of its power was as much due to a particular process in extracting it as it was dependent on a particular sort of material. Having that in mind gives us a clue, perhaps, to his well-known and enigmatic statement that burning water (alcohol) was and, at the same time, was not the fifth essence. The process was all important, and once discovered, the same procedure, he concluded, could be used to obtain the fifth essence from a variety of materials, including human blood, animals, herbs, fruits, and roots. These needed first to be "digested" (slowly dissolved through long-term cooking at low heat) with the addition of salt. Once their more subtle parts had been separated, they could be mixed with the quintessence of wine that, John instructed, was to be obtained through repeated distillation in a "circulating vessel." To prepare the fifth essence of gold, also called "incombustible oil," John advised amalgamating gold with mercury and combining the amalgam with "distilled philosophical vinegar" (probably distilled wine vinegar and something like *aqua regia,* which was a mixture of nitric and hydrochloric acids) and heating the resulting oil in the fire. Another powerful medicament, the fifth essence of antimony, was to be prepared by pulverizing antimony, steeping or soaking it in distilled philosophical vinegar, and then distilling the mixture in a gourd-shaped flask with a large mouth called a cucurbit. The resulting essence condensed, he said, into blood red drops that were indescribably sweet.

The Book of Light is more clearly metallurgical than medicinal in its purpose. Printed editions of this work appeared in the 1560s and 1570s, and these and preceding manuscripts declared that the material of the Philosophers' Stone was mercury. They described either seven or nine processes for making a red tincture that could transmute lesser metals into gold. What Rupescissa had to say in his *Book of Light* was not as well known as what he described in the *Book of the Fifth Essence*. Part of the reason is that ideas contained in the latter book also came to the attention of many people due to something unexpected that happened to the text on its way to the marketplace. At the end of the fourteenth or at the beginning of the fifteenth century, someone took parts of Rupescissa's text and combined them with excerpts from writings attributed (incorrectly, as it turns out) to the Spanish (more exactly, Catalan) theologian and philosopher Raymund Lull. The book that resulted was called *Concerning the Secrets of Nature or the Fifth Essence* and for a long time it was easy to believe that Lull was the author of everything in it. The book, incidentally, became the primary means for promoting the use of the fifth essence of wine in later medieval alchemy and medicine and was very popular in the late Renaissance as well, appearing in numerous manuscripts and in various printed editions (three at Venice and one each at Lyon, Augsburg, and somewhere undisclosed) in the first two decades of the sixteenth century alone.

So, who was Raymund Lull? Lull was a physician and philosopher who spent most of his time preparing a method of learning that, by dividing everything that was known into specific categories, made it possible for someone to know everything—a way of becoming a walking encyclopedia in all matters related to the created universe. This was the Lull that most scholars knew. However, there was also an alchemical Lull, and it is today widely agreed that none of the alchemical writings bearing his name was actually written by him. Whoever the actual author or authors of these works, the texts bearing Lull's name defined a major tradition of medieval alchemy that extends in later commentaries throughout the period of Scientific Revolution (Pereira, 1989).

Among the most famous writings in the Lull tradition were the

Testament, whose author is searching for the universal material agent of transmutation and healing, and another piece that was added as a supplement to the *Testament* as a "Codicil" (or *Codicillus*). The latter text compared human reproduction and generation to a four-stage process of alchemical labor and imagined as well an intimate physical connection (or correspondence) between the world at large (the macrocosm) and the body of man (the microcosm). It also made much of the spiritual character of the true alchemist who is inspired and enlightened through divine revelation. The underlying notion in Lullian writings is that the first matter of Genesis was mercury, a substance that continues to reside, in either subtle or more coarse forms, in every created thing—from angels and the heavenly spheres to the terrestrial elements (earth, air, fire, and water). In everything—plants, animals, minerals, metals—there was to be found a heavenly mercury, and it was through this substance that the heavenly bodies could occasion changes in generation and corruption in the things of the sublunar world. The fifth essence was itself a less pure form of the divine mercury. Because the vitality and activity of each body arose from its fifth essence, it seemed clear to Lullian alchemists that extracting a body's active principle, its fifth essence or quintessence, would be the first step in producing a powerful substance capable of transforming and perfecting other bodies. Consequently alchemy, for Lull as also for Rupescissa, was again the work of extracting the quintessence (thought of as a variety of celestial mercury) from different materials with the aim thereafter of refining and multiplying its purity and power.

Without doubt, however, the most influential of all the Lullian texts was a book called *The Book Concerning the Secrets of Nature*. If you were interested in alchemy anytime after the fourteenth century, no matter if you were Italian, French, German, or English, odds are that you would have encountered it, and sometimes in your own language. Rupescissa's ideas are, as you might have guessed, clearly present in this writing too, especially in regard to procedures making use of the fifth essence of wine (alcohol). What is different, however, is that the alcohol procedures derived from Rupescissa are now clearly directed toward producing metallic transmutations by virtue of creating a Philosophers' Stone rather than addressed to making medicines. Consistent with the real Lull's belief in an underlying mystical logic in which the order of nature matched the categories available to the mind, this alchemical Lull thought that all of alchemy could be known by reducing its parts to a kind of alphabetical code in which individual letters corresponded to alchemical principles. All you had to do was memorize certain patterns of letters and the secrets of nature and of the alchemical art would become clear. When you saw the letter "S," for instance, you knew to dissolve, purify, and recombine particular minerals and metals. It may not have been easy, but it certainly was methodical.

Someone else who influenced Lull, and who is referred to in Lull's *Testament,* is the English scholar-alchemist Roger Bacon. Bacon was yet another of the founding fathers in the tradition in alchemy that sought an elixir of life, the mother of medicines, or, as we have been calling it from later sources, a fifth essence (Newman, 1995; Pereira, 1998, 1999). There is, however, a big difference between what Bacon suggests as the material origin of this medicine and later, especially Lullian, beliefs. The Lullian author of the *Testament* declared that the alchemist had to begin his or her process with something that was already incorruptible in nature, and thus established gold and silver, viewed as the rudiments of perfection, as the appropriate materials from which a series of operations could bring about the desired elixir or Philosophers' Stone. If you wanted to end up with super-perfection, you had to use something fairly perfect to begin with. Bacon, however, looked elsewhere in the world to find a substance from which he could extract a *prima materia* or first matter. He looked to organic bodies and most expressly to human blood.

Hardly anyone creates ideas entirely out of his or her own head. Moreover, in dealing with alchemical literature of the medieval period, hardly anyone is really who they say they are. In this regard, many of Bacon's writings bear the influence of an author (some say authors) pretending to be a much respected Persian physician named Avicenna. Avicenna, who wrote in Arabic, was one of the most influential medical writers in the medieval world. Thus, the author or authors who used his name could count on attracting lots of attention to the texts they composed and could probably demand higher prices for the copies they had to sell. One such treatise, called *On the Hindering of the Accident of Old Age*, was thought for a long time to be the work of Bacon himself. Another, called *On the Soul in the Art of Alchemy*, was written in Spain sometime in the twelfth century and emphasized animal substances like blood, eggs, hair, and urine, or organic by-products like milk, cheese, and apples as the best sources for beginning a process leading to transmutation. Bacon followed suit, making use especially of suggestions found in the pretend Avicenna's text *On the Soul*. In practice his procedure for transmutation went something like this. The components of any of a variety of materials (animal, vegetable, or mineral)—but especially human blood, which was considered to contain an abundance of the fundamental stuff, the first matter found everywhere in nature—were to be separated by means of arranging the bodies into layers and then distilling the mass. The resulting purified elements needed then to be mixed with three other ingredients: a "lesser body" (in other words, the calx, or ashy powder, remaining after heating the metal one wanted to transmute), a "spirit" (mercury), and a "ferment" (most likely the calx or ash made by heating a bit of the precious metal one desired to produce more of). Precise measurements in the combination of these ingredients led to the production of an elixir that could prolong life, dispel corruption, and perfect ignoble metals (Newman, 1995).

Among medieval alchemists, Bacon is especially interesting because he has sometimes also been viewed as a heroic fixture of medieval physical science and mathematical reasoning. "His greatest title to fame," says George Sarton in his monumental *Introduction to the History of Science,* "was his vindication of the experimental spirit" (Sarton, 1927–1948, vol. 2: 953). Those who have focused on Bacon's writings have concentrated a great deal on what he called "efficient causality," in other words, the question of how actions or "species" (mechanical forces, light, and unseen influences generally) are transmitted over distances. What is going on, for instance, when a lodestone attracts a metal object, or when a room is suddenly filled with light? Because Bacon is viewed as a "scientist" or "encyclopedist" who realized the utility of physical and mathematical knowledge, it has been hard for some historians to recognize the relevance of his alchemical labors to his explorations and descriptions of nature.

Yet Bacon saw in alchemy a utility "greater than all the preceding sciences" and in one of his texts he notes that alchemy "treats the generation of things from their elements . . . Wherefore, through ignorance of this science, neither can natural philosophy . . . be known, nor the theory, and therefore neither the practice, of medicine" (Newman, 1995: 76). Those who have wanted Bacon to appear modern have indulged in a little historical transmutation of their own, arguing that passages like the one above amount to "striking proof of his scientific discernment" because there Bacon "formed a clear, though distant survey, of chemical science as an intermediate link between Aristotelian physics and the science of living bodies" (Bridges, 1897–1900; rept. 1964: lxxiv–lxxvii). As we shall also see later, people make the most amazing claims about alchemy, especially when they want it to be something else. In this regard, Bacon's alchemy gets acknowledged, but for all the wrong reasons. It was alchemy, not chemistry, that Bacon had in view, and which he believed could teach the preparation of useful things, including the production of the Philosophers' Stone.

The most useful thing of all to human beings was to find a means to prolong life. Finding the first matter, separating from it the four elements and refining and reuniting them again, would, Bacon thought, produce a perfect thing capable of bestowing its perfection on everything else. If bestowed on the human body, the recipient would enjoy health and longevity. But there were other, more accessible, medicines that could be made too; and in their preparation, mathematics, experimental science, medicine, and alchemy joined forces to compound medicaments according to particular proportions. It is to Bacon's book of medicinal antidotes that one looks to find his rules for preparing drugs according to ratios by weight. Medicines thus concocted, he thought, could help people look better and live longer; and besides ancient pharmaceuticals like *balsam, theriac,* and *benedicta* (the precise definitions of which he seems to have constantly fretted over), Bacon advised the use of other healing regimes such as song, the sight of human beauty, and, what must have seemed the most pleasant restorative to a Franciscan friar, the touch of girls (Getz, 1991: 144).

The same idea of alchemy as the best means to eliminate human suffering appears also in texts attributed to, but not written by, a medieval physician called Arnold of Villanova (ca. 1240–1311) (Pereira, 1995b). As we have seen before in regard to treatises ascribed to Avicenna and Lull, anonymous authors in the Middle Ages frequently sought to gain authority for their alchemical writings by posing as well-known and respected figures. This is also the case in regard to a text called the *Rose Garden of the Philosophers,* which many have thought to be the work of Arnold. Regardless of authorship, however, the text, which we will come back to a little later, must be considered one of the most important of the fourteenth century. Like other alchemical writings, it too extols the alchemical medicine, or elixir, as a thing possessing the most active virtue of any other remedy (Pereira, 1995a) and clearly recognizes the merits of distillation as one of the steps in obtaining the Philosophers' Stone (Telle, 1992). It is also an explicit illustration of a basic assumption held in common by almost all medieval alchemists—namely, that in doing alchemy the preparation of medicines and the transformation of metals were operations cut from the same theoretical cloth.

☀ Some readers, I suspect, are by now scratching their heads. How, you may be wondering, could people believe all this? How was it possible to conclude that alchemical promises and procedures, especially the sort that promised gold, were anything but sincerely held daydreams? I need to point out something very important. Like so many things that are very important, it is also very obvious. Alchemy came into existence and sustained itself for a long time not because it was a grand delusion but because it did make sense. It followed naturally from an intellectual context that was securely anchored to particular philosophical suppositions, religious beliefs, and social institutions. Because of the coherence of this entire set of relationships, alchemy, including the metallurgical sort, could be thought of as a rational pursuit. What we need to do is to understand how transmutational alchemy could be viewed as a perfectly reasonable and logical endeavor.

From the prevailing alchemical viewpoint of the later Middle Ages, an explanatory system that had developed out of the thinking of Aristotle and that had been expanded and embellished by Islamic authors, the various metals and minerals of the earth were thought to be composed of different amounts of two main ingredients, sulphur and mercury. According to Aristotle, the terrestrial world was composed of four elements: earth, air, fire, and water; and each element was itself composed of two separate "qualities." Earth was cold and dry; fire was hot and dry; water was cold and wet; and air was hot and wet. By exchanging one or both of their qualities, alchemists could change the elements themselves into another element. Water (cold and wet) became air (hot and wet) when the quality "cold" was exchanged for the quality "hot." Take water, heat it up, and, hey, where did it go? It changed from one element into another, of course. The point is, Aristotelian natural philosophy made elemental transmutation one of its key postulates.

And there was more. In two instances, especially, Aristotle posited the rise of intermediate substances as a result of elemental transformation. The element earth gave rise to a substance referred to as smoky earth when a shift of qualities changed earth into fire. Water, on the other hand, produced an intermediate watery vapor as the exchange of its qualities transformed it into air. The combination of smoky earth and watery vapor yielded, in Aristotle's description, the various metals and minerals of the world. Later, especially in the hands of the Arabic writers Jabir (Geber) and Rhazes, smoky earth was renamed sulphur and watery vapor also got a new name—mercury. The purity of sulphur and mercury in combination accounted for the purity and impurity of the resulting metal. Gold was the purest of all the metals in which "sulphur" was dominant, silver the purest in which "mercury" was the cardinal part.

In a later English translation of an alchemical text of the thirteenth century called *The Mirror of Alchemy*, one can read that "Alchimy is a Science, teaching how to transforme any kind of mettall into another: and that by a proper medicine, as it appeareth by many Philosophers Bookes. Alchimy therefore is a science teaching how to make and compound a certaine medicine, which is called Elixir, the which when it is cast upon mettalls or imperfect bodies, doth fully perfect them, . . . The naturall principles in the mynes are [Mercury] and Sulphur. All mettals and minerals, whereof there be [many] kinds, are begotten of these two: but I must tell you that nature alwaies intendeth and striveth to the perfection of Gold . . . For according to the puritie and impuritie of the two aforesaid principles, [Mercury] and Sulphur, pure and impure mettals are ingendered" (Linden, 1597; rept. 1992: 1–2).

The so-called sulphur-mercury theory of metals exerted a strong influence on metallurgists trying to make precious metals. But other traditions also guided their work. In this regard an important fashion in metallurgic alchemy involved reducing gold and silver to their supposed "seeds" or "souls," joining them, through distillation, with the original *prima materia,* or Mercury, in the heavens, and then recombining the purified parts (Gold, Silver, and Mercury) to produce a transforming tincture. This is the central idea of the previously mentioned *Rose Garden of the Philosophers,* a compilation of excerpts from alchemical texts whose philosophical arguments concerning the nature of metals combined with poetic illustrations of alchemical procedures to form a literary landmark of medieval alchemical theory (Telle, 1980). Through words and images, the *Rose Garden* related how *sol* (the symbol of the sun, gold, or the masculine) and *luna* (the symbol for the moon, silver, or the feminine), sometimes depicted as a king and queen, respectively, had to be dissolved in an acid bath to become one hermaphrodite body. The body is then destroyed (symbolic death), to be resurrected and further ennobled thereafter when its soul has mixed with celestial virtues. Whatever the immediate origins of this particular view of alchemical procedure, there is no doubt that it shares much in common with another text called *The Emerald Tablet of Hermes Trismegistus* (Pereira, 2000), a little composition consisting of a single paragraph derived from the ancient world and well known in the medieval era. "That which is beneath," Hermes (believed to be an ancient sage) is made to say, "is like that which is above: and that which is above, is like that which is beneath . . . Thou shalt separate the earth from the fire, the thinne from the thicke, and that gently with great discretion. It ascendeth from Earth into Heaven: and againe it descendeth into the earth, and receiveth the power of the superiours and inferiours: so shalt thou

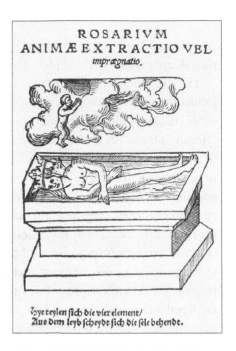

ROSARIVM
ANIMÆ EXTRACTIO VEL
imprægnatio.

Hye teylen ſich die vier element/
Aus dem leyb ſcheydt ſich die ſele behendt.

Figure 3. Processes depicting the separation and return of the spiritual part of the dissolved body of gold and silver. from the *Rosarium Philosophorum* (Frankfurt, 1550). The captions read (left to right) "Here the four elements divide [while] the soul nimbly separates from the body" and "Here the soul springs downward and revives the purified corpse."

PHILOSOPHORVM.

ANIMÆ IVBILATIO SEV
Ortus ſeu Sublimatio.

Hie ſchwingt ſich die ſele hernidder/
Vnd erquickt den gereinigten leychnam wider-

have the glorie of the whole worlde" (Linden, 1597; rept. 1992: 16) (Figure 3).

Besides the sulphur-mercury theory of the origin of metals and minerals and the view taken in the *Rose Garden* concerning the making of the Philosophers' Stone, another alchemical conviction regarding nature common in the medieval and Renaissance world is also important to understand. This is the notion that not just animals and plants but also minerals and metals were essentially active and able to grow. Some interpretations, especially those connected to Plato and his followers, asserted the existence of a vital principle in all things. In other respects the view still reflected the physical arguments of Aristotle that all things in nature sought after the perfection of their being, and this applied as much to metals as to anything else. Thus, just as a lump of coal left in the earth long enough may become a diamond, metals in the earth, when left to themselves, would naturally (over, admittedly, a very long period of time) all tend toward their respective greatest purity and perfection, namely gold or silver, dependent on their constituents. Metallurgical alchemists, then, did not attempt to impose, contrary to nature, a change of one thing into another, but sought to find a catalyst (given many names like the Philosophers' Stone or the elixir of life) that, when applied to base metals, would hurry nature along and speed up the process of perfection.

Not all alchemical theories were derived from ancient or medieval sources. Some grew out of conceptions of nature and evaluations of what constituted a primary or universal matter propounded much later, during the period of the Scientific Revolution. In this regard, some alchemists who adhered to the views of nature advanced by the sixteenth-century physician Paracelsus (1493/94–1541), sought to prepare the Philosophers' Stone from vitriol. Others, who traced their procedural lineage to an alchemist named Michael Sendivogius (1566–1636) expected to produce it from nitre. A third tradition extending well into the seventeenth century and

connected to various authors who bore much in common with medieval predecessors thought of the primary or primitive substance
as a form of mercury or quicksilver.

No matter what theoretical tradition they professed, however,
alchemists often claimed that the actual procedures they attempted
had been blessed with success. We might like to be skeptical, but
there is a sense in which such claims can be believed. An interesting discovery that one makes when reading through alchemical
writings, including the private notes and recipes of obscure adepts,
is the number of times that witnesses, including princes and members of court, testified to alchemical accomplishments, even success
in producing a tincture. Some of these reports result no doubt
from simple legerdemain. Accounts of alchemical trickery were well
known. Nevertheless, some demonstrations were made with the
strictest constraints applied; and whether the goal was a pharmaceutical elixir or a metallurgical reagent, there is no reason to doubt
the truth of claims alleging the successful completion of specific
processes (once, that is, we understand what success actually meant
within the context of alchemical observation and theory).

Alchemical labor was, after all, never a matter of a single procedure. Producing a Philosophers' Stone or Grand Elixir was fine and
good, but remained an idle performance without the means to
apply it to metals or to multiply its effectiveness. These were each
individual procedures; and it was thus entirely possible to have finished one process successfully in terms of producing expected colors, consistencies, and other effects, but then to fall short of expec-
tations in further operations (Karpenko, 1992). Moreover, while
assaying techniques existed for determining alloys, the procedure of
doubling gold—that is, of mixing gold with other ingredients (silver and copper, for instance) to increase the amount without altering too much of its color and weight—was a common practice. Although the quality of the metal would have been diminished, there
was no doubt that there was more of a precious metal at the end of

the process that there had been at the beginning. Empirical evidence could testify to the success of an alchemical procedure—if, that is, one did not look too closely.

The procedure of doubling gold and other metals led sometimes to the debasement of coins and was probably a main reason why medieval rulers occasionally found it necessary to forbid the practice of alchemy. On the other hand, alchemical projects and the confidence inspired by claims to success also contributed to the political ambitions of monarchs and princes, many of whom contributed to the support of alchemical adepts. Indeed, during the course of the thirteenth and fourteenth centuries, alchemists acquired a social identity apart from other professions and artisan activities. What most prompted the recognition of alchemy as an independent occupation was the legitimacy that alchemists obtained by being increasingly called on to serve at princely and royal courts. What brought them there in the first place had often to do with a crisis of political economy.

✳ In the period around 1300 a general lack of precious metals in Europe obstructed the expansionist plans of many territorial rulers and made their own claims to regional authority more vulnerable. To stretch their resources, some courts turned to the practical skills of assayers and alchemists, who, by alloying gold and silver with other metals, provided the court with a means of producing more coins from the usually modest amount of gold and silver at its disposal. The budgetary advantages of such processes were obvious. In England, Edward III (1312–1377) ordered that two alchemists, John le Rous and William of Dalby, be brought to him with or without their consent. The king valued their technical skills as politically significant because, he reasoned, "by that [alchemical] art, and through the making of metals of this sort, they will be able to do much good for us and for our kingdom" (Obrist, 1986: 51).

While the growth of secular monarchies was, at least in some cases, facilitated by the counterfeiting of coins, the Church found it necessary to condemn alchemy outright and labeled alchemists as nothing more than simple forgers. For territorial rulers, whose revenues from personal demesne lands were generally unequal to their political ambitions, alchemy offered a technical solution to generating wealth and extending political power. From the point of view of the Church, however, whose much larger landed holdings provided extensive wealth and at whose expense territorial rulers sought to increase their own authority, alchemy and alchemists were bad. In the evaluation of alchemy, part of the context of interest and judgment was political. Thus, in the *Romance of the Rose,* Jean de Meun, who supported the secular rights of princes, valued alchemy as a true science. "It is a notable thing," he writes, "that alchemy is a true art . . . For those who are masters of alchemy cause pure gold to be born from pure silver. They add weight and color to it with things that cost scarcely anything . . . And they deprive other metals of their forms, to change them into pure silver, by means of white liquids, penetrating and pure" (de Muen, 1971: 272–273). Around the same time, however, a papal Bull called "They promise that which they do not produce," which was proclaimed by Pope John XXII in 1317, condemned alchemists for practicing deceit and for forging coins. Alchemists, the pope declared, contradicted themselves, and presumed to carry out operations that were not in nature (Halleux, 1979: 124–125). From the religious point of view, the condemnation may also have been a rejection of the possibility that human action was capable of working on the inner processes of nature. There was also the quirky comparison between transmutation and transubstantiation (the changing of bread and wine into the body and blood of Christ) advocated by some. The later English king, Henry VI, thought that priests might be particularly good at making gold because the Catholic Mass required them to produce a literal transmutation in the celebration of the Eucharist. The English clergy was duely outraged by the comparison and refused to cooperate (Ogrinc, 1980). But there were other analogies that seemed to link Christian religious doctrine with alchemical labors. One of the most interesting German alchemical texts from the early fifteenth century is *The Book of the Sacred Trinity,* whose author is thought to have been a Franciscan monk named Ulmannus and which detailed the resemblance between the suffering, death, and resurrection of Christ and the process of creating the Philosophers' Stone (Buntz, 1971). Whatever the underlying concern, at the beginning of the fourteenth century Pope John XXII made his position absolutely clear. Alchemists were to be dealt with as criminals and their property confiscated. Where clerics were involved, they were to be stripped of their wealth and removed from whatever offices they held.

The situation was different, however, within the cultural domains established by kings and princes. In the Holy Roman Empire, France, and England, alchemists provided practical service to courts and dedicated writings, sometimes metaphorically promoting political ambitions, to enthusiastic patrons. As a technical-political option, alchemy attained its most vigorous patronage at times of intense competition between rival political factions. The Hundred Years' War provided just such an environment in England. To meet political challenges from the French and to deal with similar challenges within England as a result of the rivalry between the houses of York and Lancaster, royal decrees emphasized the need to increase the number of gold and silver coins in the king's treasury so as to satisfy the creditors of the crown. Alchemy thereafter became economic policy for Henry VI (1421–1471), who appointed royal commissions, made up of high-ranking ecclesiastical figures, royal officials, and "men learned in natural philosophy," who were to report to him about whatever they learned concerning the alchemical art. The policies of both Henry and his successor (Henry VII, 1457–1509) did much to debase English currency. At the same time, however, the apparent respect received by alchemists at the court helped to occasion an increase in popular attention both to the subject of alchemy and to its practitioners. The first alchemical works written in English verse appeared in the second half of the fifteenth century. Ripley's *Book of the Twelve Gates* and Thomas Norton's *Ordinall* were both composed at this time. Their writings were not confined to courtly readers or clever adepts, but aimed at a much wider audience. Having gained legitimacy through the court, however, alchemy could be offered as something akin to fashionable science. The works of both authors appeared with the express purpose of instructing the multitude, and even the philosophically unlearned, provided they were not evil or vicious, in the alchemical art.

❊ Nevertheless, metallurgical alchemy, even given its practical aims and utilitarian goals, remained, for some, suspicious knowledge; and although a sprinkling of interest may be found in the subject within the university, it was, as a manual art, always denied a part in the scholastic curriculum. Those who studied the "question of alchemy" from a philosophical point of view needed to be intellectually versatile and tolerant of a subject that combined matters that were practical, rational, and obscure under the heading of a single discipline. One alchemical theorist who did so was a fourteenth-century city physician named Petrus Bonus of Ferrara. His major offering was a worthy answer to the academic critics of alchemy called *The Precious Pearl.*

Bonus was not interested in describing new procedures or recipes. In fact, one of the most intriguing things about his *Precious Pearl* is that in it Bonus admits candidly that he has not devoted himself to the manual side of alchemy at all. His purpose was philosophical. What he sought to demonstrate was that alchemy was a science, with its own realm of knowledge and methods of inquiry, and that this science was nobler than others because, in part, it was based in divine revelation. To Bonus, natural knowledge was a single hierarchical structure in which particular sciences took their principles from more general realms of knowledge. In this way, he argued, alchemy could be considered a part of natural philosophy because it fell into the category of subjects that studied matter undergoing change. Alchemy inquired into the characteristics of metals and minerals. It took its general principles from the larger philosophical debates concerning those subjects and then converted those principles for use in its more limited enquiry—namely, how metals can be artificially transformed into one another. Theoretical alchemy, then, was not undisciplined. It took its lead from a more general knowledge of minerals and their characteristics and then related that knowledge more specifically to its own specific subject matter (Crisciani, 1973). So, alchemy was a part of natural philosophy after all—a branch of philosophy's many-boughed tree of knowledge. However, when it came to acquiring knowledge of transmutations, Bonus declared that this was something that could only be learned as a result of divine inspiration.

Bonus, like many others, had no problem in thinking of alchemy as both science and religion, as dependent on both reason and revelation, and as embracing both the practical and the divine. As a kind of knowledge, both reasoned as well as revealed, alchemical theory was also both dynamic and speculative. After all, what one learned through personal inspiration might take many forms. Thus, while university philosophers repeated Aristotle *ad nauseam*, alchemical theorists could take flight. In fact, some ideas about the construction of matter traditionally linked to chemical writers in the sixteenth and seventeenth centuries were already part of medieval discussions. A good example relates to the corpuscular theory of matter and the generation of metals that William Newman has recently demonstrated lies at the heart of one of the most important alchemical writings of the medieval period, a work called the *Summit of Perfection* (Newman, 1991, 2001).

According to the *Summit*, whose author was a European pre-

tending to be a well-known Arabic alchemist named Geber, the two
principles of metals, sulphur and mercury, were each composed
of tiny particles or corpuscles corresponding to the four elements
and held together in "very strong composition." In Newman's view,
it would not involve too great a leap to see in the idea of a very
strong composition—in other words, that which cemented particles together—"a kinship with the chemical bond of contemporary
chemistry." Moreover, he notes, that this corpuscular view of matter
was quite different from some ancient views that had suggested that
matter was composed of unsplitable particles (the word "atom" in
Greek means unsplitable). Bodies, as described in the *Summit*, were
composed not so that each tiny part was identical in substance with
the whole, but instead were made up of different sorts of discreet
particles held together by a powerful cohesive bond. Especially,
it was the size of particles that claimed pseudo-Geber's attention
as the *Summit* came to consider procedures of transmutation. Gold
was made of very small particles of mercury and sulphur that com-
pacted very tightly—thus its great weight. To make the Philosophers' Stone, the alchemist needed to produce increasingly minute
particles of mercury that, according to theory, could thoroughly
permeate the spaces between the sulphur and mercury particles of
a base metal and thus perfect its particulate composition. Some
things that are very old seem, with a change of theoretical attire,
to be relatively modern inventions. What is important to keep in
mind, however, is that when chemical theorists like Jean Baptiste
van Helmont and Robert Boyle (about whom we will have much
more to say later) advanced their own views of corpuscular chemistry during the period of the Scientific Revolution, they did so not
entirely out of the blue but as part of the further elaboration of a
tradition that had begun hundreds of years before (Newman, 2001:
294–300).

CHAPTER TWO

"THAT PLEASING NOVELTY":
ALCHEMY IN ARTISAN AND DAILY LIFE

Leonardo da Vinci (1452–1519) disapproved of alchemy and simultaneously awarded it great praise. In his *Treatise on Painting,* near the end of a chapter illustrating the finer points of depicting light and shade, he cast a shadow also over alchemical philosophers and artisans, calling them ingenious simpletons, fools, or imposters. Yet, if one dissects his notes on anatomy, one finds a judgment of a very different sort. There he refers to "the old alchemists" who merit infinite praise for discovering the utility of things that serve all mankind. So, which view expresses the artist's most mature and considered opinion about alchemy? Nothing easier; we should trust both impressions, of course. In the Renaissance you can love and hate alchemy at the same time. Let me tell you why.

What we might like to call chemical technology was a part of alchemy, and Leonardo himself included reference to such useful alchemy in his notebooks. He recorded the separation of gold and silver by means of nitric acid and added there a recipe for making the powerful acid *aqua regia,* or royal water. *Aqua regia,* he knew, dissolved gold, and processes for this type of alchemical metallurgy were possibly available to him already in the workshop of his teacher, Andrea del Verrocchio, who, according to the Renaissance

art historian and biographer Georgio Vasari, was an adept of the alchemical art. At his death Leonardo left behind thousands of pages
of notes, and in some of those notes found today in Milan he recorded recipes for "tinting" gold, a refining process in which gold
alloys gain the appearance of pure gold. There also Leonardo described the process of separating gold and silver by "cementation."
This was an ancient procedure in which thin sheets of a gold alloy
were alternated with layers of a mixture of salt (or, in some versions, saltpeter), brick dust (a silicate), vitriol, and perhaps some
alum, and then heated together. The silver (to use terms a bit more
familiar to us) was converted into chloride (when mixed with salt)
or nitrate (when mixed with saltpeter) and then separated away
(Reti, 1965).

Changing the appearance of metals was often part of the Renaissance artist's job description, and recipes for making yellow glass,
for making a paste that supplied a patina to bronze by dissolving
copper in nitric acid and mixing with verdigris, and for making a
beautiful yellow pigment called "the saffron of iron" were part of
Leonardo's chemical repertoire. Drawings of a distillation furnace,
known as an *athanor* (designated for the use of making nitric acid),
and a self-feeding furnace of a type that continued to be used well
into the eighteenth century entered also into the pages of his notebooks. There too Leonardo described the operations of certain natural phenomena in terms analogous to familiar physical and chemical processes in the workshop. To explain, for instance, why water
appears at the tops of mountains sprouting forth there to give
rise to streams and rivers, Leonardo argued that heat drew the
moisture inside the earth upward. Just as a kettle when heated on
the top draws water up inside it, the sun, he reasoned, heated the
tops of mountains, causing subterraneous waters to rise. Later he
argued that volcanic action rather than the heat from the sun
caused the rise of waters. Heat within the earth evaporated subterraneous lakes, he said, and the earth itself operated like a giant dis-

tilling apparatus, vaporizing sea water and condensing vapors into water again at high places (Reti, 1952).

Someone else living around the same time and who, like Leonardo, both praised and blamed alchemy was a Sienese mining engineer and metallurgist named Vanoccio Biringuccio (1480–1539). Biringuccio was a no-nonsense material man in a material world and was well acquainted with the hard realities of life and the toughness required to extract ores from the earth. Harsh political realities forced him into exile from his beloved city on two occasions. Biringuccio was simply very rich, and much of his wealth and status in Siena came from his possession of a monopoly in the manufacture of saltpeter (used for making gunpowder, and, as we have seen, an essential ingredient in processes for separating gold and silver). His knowledge of fortress design and casting armaments brought him first into the service of the cities of Parma and Venice, and then to the attention of Pope Paul III in Rome. There, near the end of his life, he was appointed director of both the papal foundry and papal artillery. Fantasy was not much required in running a foundry, and the book that Biringuccio wrote, some say dictated, called *Concerning the Making of Things by Fire (De la pirotechnia)* was in every way a straightforward practical manual of metals and metallurgy. It was published after his death in 1540 and saw four reincarnations at Venice before the end of the century. The book ridiculed alchemical daydreams; but when it came to working with metals and causing changes in them by applying alchemical techniques, it acknowledged plainly that alchemy was "the source and foundation" of many other arts.

Making gold was, in Biringuccio's view, a delusion. Alchemists simply could not, he reasoned, imitate what only nature could create. Even supposing that one could possess the basic materials from which nature composed metals, it still remained a puzzle to him how one could "receive at will the influence of the heavens, on which are dependent all inferior things . . . and also how men ever know by this art how to purify those elemental substances . . . or finally how to carry these substances to perfection as Nature does and make metals of them." To do such things required more than human skill. "I do not believe," he added, "that anyone could accomplish all this unless men were not only geniuses but also angels upon earth" (Biringuccio, 1943; rept. 1990: 37). What really annoyed Biringuccio about alchemists in other words was not what they did, but what they claimed to be able to do—particularly when they claimed to be able to surpass the operations of nature. Talk of possessing a fifth essence or Philosophers' Stone that would resuscitate the vital forces in the human body and prolong life indefinitely suffered, according to this ever-practical observer, from one fatal flaw. All those who had made such a claim were just as dead as everyone else. And yet, Biringuccio was ever alert to stating his views about transmutation and the universal medicine as an opinion about the limits of what human beings were capable of knowing. "I am discouraged," he confessed, "because I know the great weakness of our intellects . . . since we cannot know the intrinsic virtues and specific powers of things" (pp. 40–41) (Figure 4).

However Biringuccio spoke with a different voice when it came to that part of alchemy that could be understood and that stood open to view by means of practical procedures in the workshop. For Biringuccio, as for Leonardo, at least some of the processes by which nature herself produced useful things and brought about a change in the form of substances could be imitated, and indeed hastened along, by means of techniques known to the artisan. Thus, one chapter of his famous book is called "Concerning the Art of Alchemy in General," and in it our practical man practically luxuriates in singing alchemy's praises. "Besides the sweetness offered by the hope of one day possessing the rich goal that this art promises so liberally, it is surely a fine occupation, since in addition to being very useful to human need and convenience, it gives birth every day to new and splendid effects such as the extraction of medicinal sub-

Figure 4. An engraving (1698) by Christoph Weigel (1654–1725) depicts the double identity of the "alchemist" who helps nature by preparing medicines but, when desirous of gold, watches "honor, wit, money, and mercury" go up in smoke. Wellcome Library, London.

stances, colors, and perfumes, and an infinite number of compositions of things. It is known that many arts have issued solely from
it; indeed without it or its means it would have been impossible for
them to have been discovered." Thus it could be said "that this art is
the origin and foundation of many other arts, wherefore it should
be held in reverence and practiced. But he who practices it . . .
[should do so] . . . only in order to enjoy the fine fruits of its effects
and the knowledge of them, *and that pleasing novelty which it shows
to the experimenter in operation* [italics added]" (p. 337).

That "pleasing novelty," the discovery of something new, which
alchemy confers on the artisan as a result of trying out different procedures, is not only what makes things thrilling in the laboratory or workshop but is also what makes alchemy itself, as well as
other artisan activities, such an important feature of the Scientific
Revolution. In this regard processes and procedures themselves
acquire the status of artifacts, or real historical objects, when we
come to consider the influences that gave rise, in the early modern
world, to new sensibilities about how nature may be put together
and about what nature could supply when acted on by the skilled
hand of the craftsman. For Biringuccio, better living was possible
through alchemy. Making steel out of iron and giving a yellow color
to copper so as to make brass were splendid discoveries, he notes,
"for which we must praise the alchemists" (p. 70).

Where there were benefits to be discovered in nature, and where
these were open to the intellect, manual labor was needed to find
them. Alchemy, in this sense, delivered a knowledge of some of the
inner powers and potentials of things, not through revelation but
by means of the acid scars and burnt fingers that often resulted
from the hands-on manipulation of different substances. "I am
sure," Biringuccio writes, "that you understand that of all the things
created by the most High God Himself or by Nature at His com-
mand, not one—even though it be an atom or the smallest worm—
has been produced without some particular gift. And if we do not

always discern this in every thing, the cause lies in our defective vision, in our little knowledge, and in our lack of careful thought concerning the necessity of seeking hidden things. Certainly those things that have such inner powers, like herbs, fruits, roots, animals, precious stones, metals, or other stones can be understood only through oft repeated experience" (p. 114). In the practical sense, alchemy, Biringuccio seems willing to say, is to a great extent the mastery of material separation and the practical knowledge gained by means of firsthand experience in liberating the inner powers of the various parts of nature.

Like Leonardo before him, the processes of combustion and calcination were of great interest to Biringuccio. Especially it was the gain in weight displayed by metals when heated that drew his attention. Lead, for instance, when heated by itself in a furnace produced an ash (called a calx) and its weight increased by between 8 and 10 percent. "This," says Biringuccio, "is a remarkable thing when we consider that the nature of fire is to consume everything with a diminution of substance, and for this reason the quantity of weight ought to decrease, yet it is actually found to increase" (p. 58). From our present perspective, we say that a metal increases its weight and forms an oxide (in other words, undergoes calcination) when part of it absorbs, usually under the influence of heat, what the eighteenth-century French chemist Antoine Lavoisier identified as something in the air, namely, oxygen. Biringuccio did not explain the phenomenon in the same way; but when looking for an answer to why a metal should gain weight when heated, he left momentarily the din of the workshop and entered another noisy domain, this one filled with the clatter and cluck of theoretical debate and speculative natural philosophy.

To our ever-practical mining engineer, it seemed reasonable to conclude that when a metal was heated, its watery and airy parts were removed by the fire. The heating of the metal also closed the pores of the material so that no more air could get in. It was, he reasoned, these lighter parts of air and water in the body that actually counteracted the metal's heaviness. Think of it this way. When you go swimming, if you keep air in your lungs, the weight of your body won't be able to sink you. When the air in the metal was removed by heat, and no more could replace it, then, Biringuccio thought, the metal "falls back into itself like a thing abandoned and lifeless" and gains weight (p. 59). Over a century and a half later, the German physician and chemist Georg Ernst Stahl (1660–1734) theorized the existence of a material stuff in bodies that he called "phlogiston," which, some thought, accounted for combustion and calcination. One way to think about phlogiston was as a substance that possessed "negative weight." Thus bodies increased in weight when it was lost. Biringuccio considered that it was the loss and further isolation of a body's airy parts that accounted for the same phenomenon. Thus a calcined body retained more of its ponderosity "in the same way that the body of a dead animal does, which actually weighs much more than when alive. For, as is evident, the spirits that sustain life are released and, since it is not possible to understand how these can be anything but substances with the qualities of air, the body remains without the aid of that which made it lighter by lifting it up toward the sky, and the heaviest part of the element has its natural force increased and is drawn toward the center" (p. 59).

The phenomenon of calcination was of great interest to natural philosophers in the sixteenth and seventeenth centuries. Even Galileo wrote about it in a little book called *The Assayer* in 1623. But as much as metallurgical phenomena and processes were coming to the attention of philosophers, artisans too were becoming more aware of the literate and bookish side of their craft. One of the most important texts to bridge the cultural divide between scholars and craftsmen in the sixteenth century was a book about mining, metallurgy, and the chemical arts written by a German scholar named Georgius Agricola (1494–1555).

Agricola's sumptuous masterpiece, *Concerning Things of Metal* (*De re metallica*) (1556; rept. 1950) could not have appeared in more elegant attire. It was printed at Basel by the publishing house of Froben, the publisher of the works of the famous Reformation satirist Erasmus and of many other literary notables. On the one hand, its aim was to clarify the technical details and to describe the machinery of mining. It also taught the specific techniques and described the ways that instruments were to be used in assaying ores and in refining metals; and it did all this in a beautiful Latin text of which there would be at least four editions before 1657. But there was something else going on in the book and Agricola explicitly referred to it in his preface. This was the attempt to clear up the language of mining and metallurgy. Agricola was especially concerned about the proper naming of things, a concern that reflected the interest of many scholars at the time who represent a tradition of learning called Renaissance Humanism (Beretta, 1997). Although initially the product of an elite literary domain, it did not take long for Agricola's book to hit the streets; and when it did, it appeared with the same linguistic purpose, although modified for the use of lay readers.

A German translation by Philippus Bechium called *Twelve Books on Mining* was also published by Froben just a year after the appearance of the original Latin text. Significantly, Bechium made the book more user friendly by replacing the new Latin technical terms created by Agricola with terms more familiar and of more practical use to those who occasionally got their hands dirty and their bones crushed while digging out various metals and "earths." Both the Latin and German versions were based not on what ancient authorities said people did, but on what real people in the mines and in the workshop were actually observed to be doing. Agricola noted that he had "omitted all those things which I have not myself seen, or have not read or heard of from persons upon whom I can rely" (Agricola, 1912; rept. 1950: xxx–xxxi); and one of his sources, although mentioned only once in the text, was most likely Biringuccio. Ideas relating especially to the distillation of mercury and silver, making steel and glass, and purifying saltpeter, alum, salt, and vitriol through crystallization appear to have been lifted from Biringuccio's text, or to have been taken over from some common source. Whatever the connection, the books of both Agricola and Biringuccio found themselves not just sitting on the shelves of scholars but also consulted by those active in foundries and work-shops all over Europe. The first of three French translations of Biringuccio's *Concerning the Making of Things by Fire* appeared in 1556, and an Italian translation of Agricola's volume could be ac-quired just slightly later, in 1563.

These were the grand texts of mining and metallurgy in the six-teenth century. However other books, also revealing knowledge of metals and chemical processes, could be easily obtained—although, in general, they existed in fewer numbers and lived shortened lives. Quite simply, these books were usually read to pieces, were occa-sionally burnt at the edges, and were sometimes warped and disfig-ured with spills and splatters. The books that had the greatest direct influence on people's experience were usually opened and con-sulted so often that they never looked great on the shelf—if they could be found in one piece at all. These books were cheaper and less attractive than the tomes of Agricola and Biringuccio, but lots of tradesmen read them. In fact, more people were reading in cen-tral Europe in the sixteenth century than ever before, and a good deal of what they read, or had read to them, fell into the category of household alchemy.

✳ In 1569, the Frankfurt bookseller Michel Harder came home from the Lenten book fair a happy man. He had sold 5,900 books at the fair, and one of his biggest sellers was a *Book of Household Medi-cines,* a book of recipes and instructions in German for preparing medicines at home. He had sold 227 copies of this book alone

(Engelsing, 1973: 26). Harder was not the only merchant to cele-brate business success of this kind. There was money to be made in selling books, especially if one could figure out what people wanted to read. If you had lived in Hamburg around 1600 you would have found it hard to be out of the sight of a bookseller. There the num-ber of book and music dealers is estimated to have been around 4,000. That's about 10 percent of the city's population! More than a half century ago, Rudolf Hirsch prepared a checklist of early printed books relating to alchemy and chemistry in an attempt to establish a sense of "the attitude and state of learning of the reading public" and "the degree of the diffusion of [alchemical/chemical] knowledge." Of the alchemical and chemical books published be-tween 1469 and 1536, he found that the most significant group comprised texts that were intended for the craftsman, many written in the vernacular or local language, and many of these written in German (Hirsch, 1950). Craft alchemy as well as household al-chemy had almost instantly become part of a popular publishing milieu.

Studies of literacy in Europe in the late medieval and early Re-naissance periods relate the general expansion of reading ability to the growth of cities, the requirements of new merchant enter-prises, and the developing customs and manners of courtly life. Of the books published before 1500 in German-speaking areas, about 6 percent, an estimated 3.2 million copies, were printed in the vernacular (Engelsing, 1973: 16). Between 1501 and 1520, an esti-mated 34.9 million copies of books (34,900 editions at approxi-mately 1,000 copies per edition) were printed in Europe, a third of them in Germany. Books, of course, are not very useful unless one can read them. Among those learning to read, artisans began to make up more and more of the literate public. It was just good for business. In London, for instance, the statutes of the goldsmiths from 1478 and 1490 required that their members be able to read. In France, in Montpellier between 1580 and 1590, 63 percent of municipal artisans were able to read and write. By the end of the six-teenth century, the demand for literacy in all ranks of society be-came increasingly urgent. The message is clear in a German poem of 1581 called *A Sincere and Lovely Description of the Art of Writing* (p. 33), which emphasized the social need to read and write among princes, prelates, and other potentates, and noted also its impor-tance to the professions and artisans:

> The doctor for his pharmacy
> Also needs literacy.
> To the craftsmen's profitability
> Writing serves much utility
> Thus among us in German lands
> A certain saying has come to stand:
> Someone is just half a man,
> Who neither reading nor writing can.

Vernacular books connected to craft and household alchemy could have gotten lost in the period that also saw the publication of books by Copernicus, Kepler, and Galileo. And yet, popular manu-als made a considerable impact in the lives and occupations of a large part of Europe's literate population. Two examples of vernac-ular texts that describe practical alchemical processes and that were meant to be read at home or in the workshop can help us under-stand what real presence they had. One sort we will look at com-prises a family of medical literature that was written, in part, for professional communities, but had as well a public presence, and was often consulted by people who desired to make medicines at home or who needed pharmaceutical assistance while on the road. The second type of vernacular literature, from which we can ad-dress only a specific specimen, relates to a whole class of books that was wildly popular in the sixteenth century known as "books of secrets."

While professional apothecaries frequently turned to dispensaries and pharmacopoeias written in Latin for lists of drugs and descriptions of their properties, preparations, and use, there existed alongside the official inventories a growing vernacular literature in the early modern era rich with references to the making of medicines. Some books were meant for those in charge of hearth and home while others aimed at the further education of artisan pharmacists. Sometimes the intent of the book was to describe remedies that only an accomplished few could make, and these functioned as an advertisement for specialized practitioners. More often, however, publishers sought a wider audience and sold books of different sorts, some directing how to prepare medicaments made from herbs, others describing more complex remedies, including those requiring alchemical expertise. In this regard, the story of a particular book, written in German, that focused on the preparation of chemical medicines and that, through re-editions, commentary, and controversy, continued to influence almost the entire seventeenth century, begins in the German city of Coburg. There a distiller at the court of Saxony and Brandenburg named Johann Popp (also Poppe or Poppius) published, in 1617, a book called *Chemical Medicine*. Popp's interests were partly medicinal, partly alchemical, and partly astrological. There was nothing unusual about that. He simply followed a medieval and Renaissance tradition that assumed that specific parts of the created universe were intimately connected to other parts, and that the powers of planets charged up related objects on earth with their own specific virtues. Each star or planet, he reasoned, possessed its own special nature, characteristic, and effect that, by means of its rays, impressed those same attributes and potencies into corresponding growing things in the terrestrial realm, even metals and minerals.

As with Raymund Lull, Arnold of Villanova, Roger Bacon, and others, Popp defined alchemy as the separation of the gross from the subtle and spiritual parts of nature, and he viewed a body's spiritual, fifth essence (the part derived from the stars) as the source of all medical cures. The proper physician had to understand "the anatomy of essences" so as to work with nature, and not against her, in attempting to restore the body to health. Thus, much of what Popp wrote about in his *Chemical Medicine* and in a supplementary "Guide" published ten years later dealt with the description of processes for extracting fifth essences from plants, animals, minerals, and metals (including gold, silver, and mercury) and for the preparation of medicinal waters and spirits. Popp wrote for the good of the public and had a particular interest in reaching vernacular reading doctors. The last installment of his three-part text appeared in 1627, but this was by no means the last opportunity readers would have to get acquainted with his ideas.

People were eager to read Popp's book. Because the book was such a good seller, others sought to cash in on its popularity, and did so while at the same time advancing their own opinions about the making of medicines. In fact, being critical of what Popp had described, and replacing or augmenting Popp's initial recipes with new formulas for making chemical medicaments, turned out to be an excellent sales strategy because even those who possessed Popp's original version might have wanted to get their hands on a corrected and enlarged rendition. In this regard, notes and commentaries based on Popp's *Chemical Medicine* were published in two parts in 1638 and 1639 by a physician at Leipzig named Johann Agricola (b. 1589). Although hindered in his full examination of Popp's pharmaceutical recipes by the ravages of war, Agricola nevertheless felt confident enough to make observations and judgments concerning Popp's processes on the basis of his own extensive experience. There was as well another sort of appraisal that prompted his decision to publish. The need for chemical medicines was all the greater in his own day, Agricola believed, inasmuch as illnesses were now more severe than they had been in the time of the ancients. Older remedies were thus often ineffective in combating present-day maladies. New remedies, more powerful than the illnesses that had emerged, needed to be found, and, he concluded, the preparation of such medicines (corresponding to the increased severity of disease) could only be learned in the school of the chemical arts. Finding new medicines was a difficult manual task, and it was made even more troublesome by the fact that learned medical practitioners appeared to disregard their responsibility in the effort. Who could blame them, Agricola asks with a voice full of contempt, for who among them would want to stick his tender hands and fingers stacked with rings into the ashes (Figure 5).

Agricola did not agree with Popp about every process, and there were those who did not agree with Agricola either. One who did not was a former court apothecary in eastern Germany named Georg Detharding. According to him, Agricola's notes and commentaries had done more harm than good to medical chemistry. Speaking "to all the lovers of true, non-counterfeit *chymia*," Detharding put Agricola's procedures to the test in a book called *The Assay Furnace of Chemistry* and found a way, in so doing, to engage again Popp's original processes while, of course, encouraging interested readers to buy yet another book. Even at this point, however, Agricola's notes and commentaries to Popp's initial text survived to see yet another reincarnation. This one appeared near the end of the century, in 1686, written by a physician named Johann Helfrich Jüngken (1648–1726). Jüngken also wanted to correct errors in the text, but was most interested in offering doctors, by means of clear instructions, ways in which they might "grasp the coals," in other words, make medicines themselves. In his introduction, Jüngken further noted something of the popularity of the Agricola-Popp recipes. He described Agricola as a much-loved man known almost everywhere and about whose cures and chemical knowledge many still alive could testify. The recipes and procedures recorded by him had been much desired but were no longer to be found anywhere except in certain well-established libraries, doubtless because other copies had been read to shreds at home and in the pharmacy. Because the

Figure 5. Title page of Johann Agricola's commentary on the chemical medicine of Johann Popp with symbolic references to themes of death and resurrection in the production of medicines and the transmutation of metals. *Erster [ander] Theil . . . commentariorum, notarum, observationum . . . in Johannis Poppii Chymische Medicin* (Leipzig, 1638–1639). University of Wisconsin Library.

demand for the book was so great, Jüngken had decided to re-edit Agricola's notes, making corrections where necessary and adding, in turn, his own commentary (Moran, 1996a).

Significantly, Jüngken dedicated his volume not to a prince but to a princess—the princess Elizabetha Amalia Magdalena of Bavaria. He thought to do so partly out of a sense of duty, having recently been named to the position of provincial physician in Bavarian lands, but also, as he says, because "noble chemistry is not unsuited to the feminine sex" and because the princess had demonstrated a general inclination to the subject. Court pharmacies were often derivatives of court kitchens, and in fact the line between kitchen and apothecary was not always clearly defined. In this case what is important is that a long tradition of preparing chemical medicines had found a way not only to be associated with the court but had also become a vernacular subject suitable to women. In fact traditional lines of social and intellectual space get substantially erased at this point in Jüngken's book. Written in German, dedicated to a princess, Jüngken's text begins with a poem composed by a university professor who was also one of the court's personal physicians. The poem crossed all sorts of social boundaries, linking together the work of scholars and laymen, physicians and alchemists, and pronouncing that a tradition partly alchemical and occult had joined with other, more practical and economically successful ventures, to become an ever-increasing treasure for the good of humankind.

In German-speaking lands, books of medicines were addressed to "house fathers" and "house mothers" of all social orders, although, as we will see, medicines of the more complicated chemical sort (as opposed to simple herbal remedies) were most often reserved for the wealthier ranks. One well-known author, Lorenz Fries, wrote for "lay people of every stripe" who because of material and/or geographic necessity had in times of illness to substitute for physicians and apothecaries and take refuge in self-medication. Lay books of medicine sometimes contained a vernacular wisdom fixed already in the late Middle Ages. Such a text was *The Little Book of Medical Practice* (1541) written by Walther Hermann Ryff (ca. 1500–1548), which copied large parts of an earlier thirteenth-century pharmacopoeia. However, other books of medicines introduced a more up-to-the-minute professional literature to lay readers and sometimes described procedures known previously only to experts.

Concerns for the living conditions of the "poor common man," and the limited possibilities of "poor and common people" to procure required medicinal ingredients prompted authors to enter the marketplace with books promising to communicate simplified cures instead of complicated, more expensive procedures. An important book in this regard is the *Public and Private Apothecary* (1622), written by a doctor of medicine in Silesia named Martin Pansa. What makes Pansa's volume so interesting is the way that certain medicaments are reserved for certain social classes, with chemical medicaments appointed for use among the upper social echelon. In other words, chemical medicines had become specialized medical commodities appropriate to social and economic advantage.

Pansa described his text as a "city, court, and house apothecary" in which one could learn about the types of medicines that should properly constitute municipal, princely, and noble pharmacies. In addition, he intended to make known to persons of wealth the latest, most valuable medicines, especially those prepared by chemists. The medicines described for this section of society are of the "watery" sort, that is, distilled waters, spirits, and oils, as well as balsams, juices, tinctures, extracts, and essences. For household use, Pansa added a "poor man's treasury," "a list of medicines for the common man which requires very little or no cost" and whose ingredients were available to anyone and could be used to combat

most illnesses. Recipes for the poor usually involved a single ingredient, whereas those advised for the high born or well off were
compositions of several materials requiring multiple processes and
considerably more time and effort to prepare (Telle, 1982).

As we have seen, the art of distillation in the sixteenth century
had, by and large, not yet come into the hands of university-
educated philosophers and physicians. The production of distilled
waters was predominantly an artisan activity. Brunschwig described
procedures derived from learned alchemical authorities like
Avicenna and Arnold of Villanova, but he also noted that his instructions were meant for "lay people, men as well as women." He
was, of course, not alone in offering a knowledge of distillation
techniques to a wide social spectrum. Other texts as well extended
this aspect of the alchemical arts to the vernacular reading public. A
popular book in German was *A Pharmacy for the Common Man*
(1529), which printers liked to offer together with another text
called *On Distilled Waters*, first published in 1476. One could buy
collections of recipes taken from various authors and sometimes
find distillation procedures put together with herbals so that, if you
were of modest means, you could produce distillates from common
plants. Books were addressed to "alchemists, barbers, apothecaries,
and households" or to "rich and poor, learned and unlearned" and
in this way offered themselves to portions of society varying widely
in educational advantage.

Medicinal distillation was keeping company with a great many
people in the late sixteenth and early seventeenth centuries, and
one could hardly find a better indication of how distilling medicines could be thought of as a form of popular alchemy than a book
that promised *The Best Part of Distillation and Medicine*, written
in 1623 by a German physician named Conrad Khunrath. In the
text Khunrath wrote that everyone, male and female, priest or lay
person, rich or poor, got sick, and it was for the relief of human suf-

fering, no matter what one's position in the social hierarchy, that God had placed helpful medicines in animals, vegetables, and minerals. However, the various parts of creation needed to be acted on so their medicines might be found. Thus, in addition, God revealed the art of *chymia*, which is also called the art of separation, or alchemy, so that, by its means, one could obtain the effective powers and virtues in things by separating away the impurities and poisons lurking within the substances of nature from their subtle and beneficial parts. Through alchemy, one learned how to prepare wondrous medicines that, because of their delicacy, penetrated the outer members of the body better than coarse, unseparated remedies and pierced directly to the body's affected region.

There was nothing new about this art of separation, Khunrath observed. It had been well known to ancient physicians, Arabs, Greeks, and Latins, and had been held in high regard by church fathers like Augustine and other theologians who understood that its foundations were to be found in Scripture. All these recognized that the true art of alchemy was the true philosophy of the wise, having not only great utility as it taught how to melt metals, to separate, and otherwise to make useful things, but instructed as well how to sublime and distill plants and animals, and to make from them life-sustaining medicines. The art showed how to extract powers from animal, vegetable, and mineral things, how to distill their subtle oils, and how to prepare their salts. Beyond that, it explained how to separate their pure from their impure parts and bring into being by art that which nature promised but had not produced. And one thing more, *anyone* could do it. The art of separation was open to everyone because alchemy had always been in great part a popular technology, a doing and making of something out of nature's ingredients. In the process one learned how things acted on and suffered one another and became alert to the relationship between actions and results without necessarily insisting on a mechanism or structure to interpret them. Doing did not require belief, but it did result in practical knowledge.

✳ Against the background of a growing demand for vernacular books suitable to domestic and craft environments, one recent author, William Eamon, has identified, and in great part reconstructed, what he calls a "book of secrets tradition" and has, in a clear and brilliant discussion, shown how a variety of handbooks bringing useful, practical information to the public began to take on the appearance of scientific and technical encyclopedias (Eamon, 1994). For our purposes four tracts examined by Eamon, collectively known as the *Little Books of Tricks,* offer a good example of the way in which books of secrets combined with household and craft alchemy to promote a routine engagement with procedures and processes that themselves contributed to the production of technical knowledge through everyday experience. The books of tricks were, in fact, technical manuals that offered easy instructions to novices and taught more sophisticated techniques to those already trained in the crafts.

Numerous editions had already appeared before 1533, but the publication of a new edition of the first of the tracts at Frankfurt in 1535 called *The Proper Use of Alchemy* was a direct attempt to make alchemical techniques part of the legitimate, for-profit daily routine of artisan entrepreneurs. The publisher of these alchemical methods had especially goldsmiths and jewelers in mind, but other craftsmen would also have been interested in the book's recipes, such as how to make artificial amber or artificial pearls, how to separate gold from copper, or how to soften gold so that it could be worked in a solid-cold state more easily. The second of the series of manuals contained recipes for making ink and colors and was aimed principally at those involved in illuminating books and manuscripts. Thereafter you could learn how to dye fabric and remove stains. In the last manual of the series, you could find processes for hardening steel and iron, and for soldering, etching, and coloring metals. This final tract, says Eamon, "was the first printed work on the technology of iron and steel," and "represents the cumulative, practical experience of generations of medieval craftsmen which, for the first time, was now being revealed to the general public" (Eamon, 1984: 121).

The fourth manual in the series represents the combination of craft skills and alchemical traditions in other ways as well, particularly in that part of the text that deals with preparing solutions that were to be used for hardening steel tools. According to the manual, files should be quenched in linseed oil or the blood of a he-goat, while cutting tools managed better in a bath made of the juice of radishes, horseradish, earthworms, cockchafer grubs, and he-goat's blood. Drill bits, on the other hand, required a man's urine along with other ingredients. These interests were by no means inconsequential, and not just a few artists made use of such recipes in the service of demanding patrons. At the Medici court in Florence, for instance, an interest in fashioning sculpted objects from hard stone, like porphyry, gave rise to collaborations bringing together the skills of those acquainted with alchemy and botany for the purpose of producing a perfect tempering agent that would make steel tools so hard that they could turn even the toughest stones into artistic creations. Where the Medici prince Francesco liked to visit court workshops and became fascinated by technical expertise and "books of secrets," his successor Ferdinando chose to patronize efforts that could amaze onlookers by reflecting an artist's ability to work his will on the most resistant materials. What was really on display, and what any image-conscious court visitor would have had difficulty missing, was not altogether an object, however, but a performance. Fashioning something wonderful from the most unyielding stuff was a not-so-subtle metaphor for the power of the prince whose personal hardness and spiritual control could overcome any opposition to the reshaping of a reluctant political environment (Butters, 1966). Alchemists, metallurgists, and artists had combined their skills to create both beautiful shapes and political symbols.

Books of secrets were already known in the Middle Ages. An Arabic work pretending to be a work by Aristotle called *The Secret of Secrets* found its way into Latin in the twelfth century and another text, a book on the secrets of joining things together, appeared also in medieval manuscripts. In these altogether-practical texts, there was no distinction made between experiment and experience; both terms denoted empirically based, reproducible technical knowledge. With the expansion of such books through print, however, scholars began to consult them more and more, and the stage was set for the recognition among natural philosophers that craft secrets amounted to a legitimate field of inquiry for scientific investigation. While books of tricks appealed most to artisan readers, those with learned credentials who consulted them and drew on the craft marvels that they disclosed began to publish their own books, announcing themselves as "professors of secrets." In these, books of secrets appeared as collections of scientific experiments and were aimed at a different reading public, especially at persons with wealth and leisure interested in amusing themselves through acquaintance with novelties that resulted from an adeptness in manipulating parts of nature. For such professors of secrets and virtuosi like Girolamo Ruscelli (1504–1566) and Leonardo Fioravanti (1518–1588), "curiosity," Eamon says, "was just as strong a motivation to examine secrets as utility." Ruscelli went further than others, however, describing in a book called *New Secrets* (1567) a design for an academy at Naples called the *Accademia Segreta* (Academy of Secrets), which acted as an experimental clearing house for those things reported in other books. His own book contained 1,245 recipes, each, he exulted, experimented on three times (Eamon, 1984: 129ff; 1994: 134ff).

✳ If you were an angel and wanted to have sex with mortal women, completely against the law of God, of course, what would you be willing to give in return? According to the apocryphal Old Testament *Book of Enoch* (probably written in the second-century BCE), one member of an especially lecherous group of angels taught earth women the alchemical crafts of metallurgy, dyeing, and the making of cosmetics and precious stones. "And they took wives unto themselves, and everyone chose one woman for himself, and they began to go unto them . . . And Azaz'el showed to their chosen ones bracelets, decorations, (shadowing of the eye) with antimony, ornamentation, the beautifying of the eyelids, all kinds of precious stones, and all coloring tinctures and alchemy. And there were many wicked ones and they committed adultery and erred, and all their conduct became corrupt" (1 Enoch 7,1; 8, 1–3; Patai, 1994: 21). God, incidentally, got really angry at this flagrant violation of the law of heaven, and in a classic case of blaming the victim thought first to destroy the entire earth. Luckily, divine wrath found a focus closer to home. The earth was saved, and the most offending angel, Azaz'el, suffered the worst consequences, probably wishing he had never been created. Nevertheless, the damage was done. Women knew alchemy. The story is of course a myth, but it does draw our attention to the fact that, when discussing alchemy of the household and artisan sort, many practitioners were women. Indeed, the role of women in the preparation of alchemical agents and their familiarity with various types of process is significant in relating popular or household alchemy to the shifting of experiences that make up the Scientific Revolution, even when those experiences, like many women themselves, were most often confined to the home.

Certainly one of the most interesting books of secrets published frequently in the sixteenth and seventeenth centuries was that of an otherwise-unknown Italian female author named Isabella Cortese. It is of course entirely possible that Isabella was a man, but there is also no reason why we should not take the reputed gender as authentic. Isabella was a female alchemist with secrets to sell, both of the grandiose sort and of the household-bedroom variety. About her life she confided only that a thirty-year study of alchemy and a thorough reading of the works of famous philosophers had resulted in nothing and had only promoted the likelihood of an early death. Then, however, she discovered secrets on her own, through her own processes, and these had brought her back to health and restored her fortune. Her book, *The Secrets of Lady Isabella Cortese* (1561), assuredly helped in this latter respect. Not only did it put money in the pockets of the author but also into those of several publishers, as the work passed through at least eleven editions between 1561 and 1677 (all published in Venice) and became the basis for a German translation published twice in Hamburg and once in Frankfurt near the end of the sixteenth century.

Cortese described her purpose as pure, professing a compassion for humanity while instructing readers to reject "grand masters." Prudence was required of those in possession of her techniques; and even though she went public with a book about secrets she advised, indeed insisted on, well, secrecy. After disdaining the works of Geber, Lull, and Arnold of Villanova as "soothing stories," Isabella laid out ten commandments, admonishing her readers (once they knew how to prepare pure gold and silver, to build vessels, and to make correct use of the fire) never to divulge their art, nor let anyone enter into their workplace. In other words, the alchemical merchandise that Isabella had to sell only remained profitable if people refrained from passing her information along to others. Keeping secrets secret while selling books of secrets was the author's real secret, which was also the secret of the marketplace.

Isabella's book was little in size but big on advice. One recipe instructed on how to join metaphorically body, soul, and spirit in a process combining fixed camphor, quicksilver, and sulphur so as to create a universal medicine. But to this Isabella also added instructions on how to mix glues and polishes, and on how to make soaps and cosmetics. One could learn how to make gold or how to concoct a toothpaste made from white wine. There was a face cream to make the skin white and velvety; and for those getting desperate, Cortese disclosed a mixture of quail testicles, large winged ants, oriental amber, musk, and an oil made from elder and storax designed to "straighten out the [male] member." It may not be easy to regard a face cream as an alchemical recipe or to think that alchemy was involved in promising sex for life, yet Cortese was sure that they all belonged to the same category—the production by art, and through secrets, of what nature herself had not delivered.

If Isabella had a workshop, it might well have also been a kitchen, and here too one could find ample signs of alchemy. Two of the most popular English books on cookery in the early seventeenth century were those of Sir Hugh Plat and Gervase Markham (1568?–1637). Plat's works, *The Jewell House of Art and Nature* (1594) and *Delightes for Ladies* (1609), gave advice on domestic duties and, in this regard, drew a line, based on gender, between inner and outer household domains. Cooking and distilling belonged to the province of women, while gardening and husbandry were to be attended to by men. The secrets of the inner realm included processes of distillation; and within the domestic context, distillation not only became a technology of better living but also brought down to earth the more arcane features of chemical medicine. Plat openly acknowledged that women imitated in the home the practical parts of mystical philosophies discussed in the circles of savants. Markham, on the other hand, so bound cookery with pharmacy that in advising on how to become a "complete woman," he delayed discussing cookery, which he considered a form of outward knowledge, in favor of instructing first on one of the "inward virtues" of every housewife, namely, the knowledge of preparing medicines. Later, in the same text, so as to help "sort her mind to the understanding of . . . housewifely secrets," he directed that the English

Figure 6. Women and distillation, from the title page of Hannah Wooley's *The Accomplished Ladies' Delight in Preserving, Physick, Beautifying and Cookery* (London, 1675). University of Wisconsin Library.

housewife "furnish herself of very good stills, for the distillation of all kinds of waters, which stills would be either of tin or sweet earth; and in them she shall distil all sorts of waters meet for the health of her household" (Markham, 1615; rept. 1986: 125).

Later in the century, with a new round of cookery books, there appeared *The Queen's Closet Opened* (1655) revealing the recipes of Queen Henrietta Maria. The book went through five editions in the 1660s, 1670s, and 1680s; and with the second edition, a new paragraph appeared in the preface declaring that what had previously been referred to as recipes "we shall now rather call Experiments" (McKee, 1998). People liked that word "experiment" in the mid-seventeenth century. It implied real science, and some brought the term home where no one was as yet insisting that impeccable distinctions be made between chemistry, alchemy, and cookery and where all three could, as they had for centuries before, continue to be warmed by the kitchen stove (Figure 6).

One person, a woman by more accounts than her own, who probably knew her way around a kitchen but who preferred to use a well-equipped laboratory when preparing medicinal remedies, many of them chemical medicines, was the later seventeenth century French writer Marie Meurdrac. In 1666 she published a remarkable book called *Benevolent and Easy Chemistry, in Behalf of Women,* a text that saw two additional French editions before the end of the 1680s and an Italian translation published at Venice in 1682. Meurdrac was self-taught and refused to remain silent about that which she knew, swimming resolutely against the tide in the long debate about women's education in France. She declared in the preface of her text "that the mind has no sex, and if the minds of women were cultivated like those of men, and if we employed as much time and money in their instruction, they could become their equal" (Meurdrac, 1666; rept. 1999). She was also a chemical practitioner and, although the lack of a professional title or license prevented her from selling her medicines publicly, she was nevertheless able to give her remedies to the poor and to make use of her laboratory for the instruction of other women.

On the one hand, Meurdrac's work falls into the tradition of the books of secrets. She describes the preparation of compound remedies and addresses one part of her text specifically to women, treating there "all the things that may conserve and increase beauty." Yet Meurdrac is clearly not just relying on an oral tradition or on books of recipes for her chemical knowledge. Her book tells us that she was aware of the major chemical writers of her day and was also clearly knowledgeable about medieval traditions of distillation alchemy, especially the literature focused on the extraction of fifth essences influenced by Rupescissa and Lull. Following from these traditions, Meurdrac considered mercury to be the "spirit of life" that the process of separation from terrestrial impurities could make more powerful so as to allow it to penetrate into the deepest parts of the body (Tosi, 2001).

After first treating the definitions of chemical principles in her text, Meurdrac described the instruments of the workshop and treated as well the preparation of tinctures, waters, essences, and salts with her main emphasis on distillation. In the fourth part of the work, concerning minerals and metals, she elected to pass over operations involving gold and silver in favor of describing more useful medicines, giving instruction on a variety of preparations including the spirits of vitriol, nitre, sea salt, and sulphur, the essence of amber, the tincture of coral, and the "crocus of antimony" (antimony sulphide). Such chemical medicaments had long been described in other books, of course, and Meurdrac was certainly aware of the controversies surrounding the use of medicaments made from minerals and metals that those books had aroused among French physicians. Her book therefore sought a conciliatory middle ground. With reference to the Bible, she acknowledged that many of the things found in nature took part in the chastisement of mankind and thus needed to be acted on through the knowledge of specific techniques in order for them to acquire a beneficial, healing effect. Such things, she wrote, furnished very salutary remedies, but only as a result of preparing them exactly and only when used in small quantities for rebellious and persistent illnesses (Meurdrac, 1666; rept. 1999: 129).

✻ Science is usually considered a cognitive realm; I suggest that it is also an existential one, that is, one made up of numerous creative experiences at home, in the workshop, as well as in the library. Scientific revolution is usually connected to a kind of ideology of genius. I suggest that it is part of a larger reality, the "strung-along and flowing sort of reality which we finite beings swim in," the sometimes-confusing experience of a reality "where things happen" (James, 1909; rept. 1996: 212–213). This allows science and changes within science to be more than a matter of intellectualism and to admit as relevant experiences of many kinds—all part of the slow shifting of perspective that allows things to appear from a new center of interest. Among others, William Eamon has noted the role played by popular traditions in the arts that searched for the secrets of nature and in the process offered new types of experience around which could crystallize new perspectives of nature. What flourished in this milieu was a confidence in the ability of empirical inquiry, experiment, and the production of effects through human agency to extract that which was hidden in nature and, by so doing, to construct a better foundation for natural speculation. Philosophy and the arts would have to cooperate in discovering the theoretical bases of useful knowledge, and in this endeavor the seventeenth-century English philosopher Francis Bacon was not reluctant to include among revered artisan traditions the long experience of practical alchemy. "It was not ill said by the alchemists, 'That Vulcan is a second nature, and imitates that dexterously and compendiously which nature works circuitously and in length of time.' Why therefore should we not divide Natural philosophy into two parts, the mine and the furnace; and make two professions or occupations of natural philosophers, some to be miners and some to be smiths?" (quoted in Eamon, 1984: 138).

PARACELSUS AND THE "PARACELSIANS":
NATURAL RELATIONSHIPS AND
SEPARATION AS CREATION

Alchemy, of course, was not just a craft. For a very long time it also flourished as an important part of natural philosophy and religion. There is nothing at all bizarre or unique in this. The medieval and Renaissance worlds simply embraced alchemy, as they did other forms of natural inquiry, such as exploring the heavens or prying into the human body, as inherently devotional activities. Later, when alchemy combined with esoteric traditions in the eighteenth and nineteenth centuries, spiritual interpretations of the alchemist's efforts helped to foster a popular belief that preparation of the Philosophers' Stone included the spiritual preparation of the alchemist himself. That view—in other words, the notion that personal transformation is somehow connected with doing alchemy—has lingered into the modern era. As many readers may know, it became a prominent feature of psychology when the psychoanalyst Carl Jung argued that alchemical imagery was a product of a universal or "collective" unconscious and could be read as revealing stages of individual psychic growth.

In the period of the Scientific Revolution, the traditions that supported the spiritual side of alchemy were already well established. While some were connected to the Bible and to Christian cosmology, others were linked to ancient philosophies that sought ways of unifying matter and spirit. One tradition especially, linked to writings reputed to have been written by an ancient sage known as Hermes Trismegistus (Hermes the thrice blessed), served to establish alchemy as a sacred and magical endeavor. In the 1960s, the historian Frances Yates argued that these texts, especially two called the *Aesclepius* and *Picatrix*, were responsible for a barely explored cultural legacy of magical knowledge that she called the "hermetic tradition" (Yates, 1964). Scholars in the seventeenth century determined that the texts of Hermes had actually been written in the early Christian era. Nevertheless, Renaissance writers knew nothing of the deception and accepted Hermes as real and very, very old. The antiquity of Hermes was important because many believed that the most ancient authors stood closest to an originally pure wisdom divulged by God to a privileged few at the outset of human history.

The rebirth of hermetism began in 1460 when a monk collecting ancient Greek manuscripts for the Italian prince Cosimo de Medici appeared back in Florence with an incomplete collection of recently discovered hermetic treatises. The job of translating the ancient writings fell to a scholar named Marsilio Ficino (1433–1499), who until then had been busy translating the texts of Plato. Ficino's translation and commentary, called the *Pimander* after the first treatise in the collection, appeared in print in 1471. The translations created a deluge of interest because what people read about when they sat down with a copy of the *Pimander* was magic.

All of physical creation, the hermetic writings explained, stood on an astrological foundation in which the celestial bodies, sometimes through the mediation of a cosmic spirit *(spiritus mundi)*, provided a link between God and terrestrial things, including human beings. Thus the planets, which included the sun and moon, as well as the zodiacal signs, influenced earthly matter by infusing their divine virtues into everything in the world. According to hermetic reasoning, people possessed divine souls; but, as physical beings, they were nevertheless subject to the stars. Human divinity was, so to speak, smothered in material obsessions. If, however, a person could ever get free from such excessive affection for physical things and purify the soul in the process, that person might not only obtain a knowledge of God but regain his or her true, unblemished divine nature—and further, such a transformation would allow the person to become a *magus*, or magician. A magician possessed an intimate understanding of the operations of nature and knew how to manipulate natural processes so as to direct the powers and virtues of earthly things for good purposes. Maintaining human health by fashioning processes that could either enhance or counteract the celestial torrent of helpful and not-so-helpful influences flooding the body and spirit was generally recognized as a particularly useful and beneficial aim.

The uncorrupted knowledge of nature and nature's powers might be achieved through reading ancient texts, but it could also come to light through direct searching and inquiry into the things of the terrestrial world. Magic and empiricism, while strange bedfellows in the house of modernity, got along quite well in the "enchanted garden" of the early modern estate. Indeed, as we will see, an experimental approach to nature had much in common with the assumptions of hermetic philosophy. Some years ago, the neurologist and psychiatrist Victor Frankl referred to the fact that a cylinder, cone, and sphere each cast a circular two-dimensional shadow (Frankl, 1988: 22–25). He called the phenomenon "dimensional ontology"; and what he meant to demonstrate by this simple truth with a fancy name was this—those things that look so clear and distinct in one context can seem altogether ambiguous in another. Magic and experiment may be very unlike and clearly distinct from the perspective of contemporary science, but the shadows they cast on the walls of the Scientific Revolution are nevertheless very much

alike. And one place to look for an illustration of how those shadows overlapped is in the writings of a Swiss-German physician,
natural philosopher, and alchemist with an impossible-sounding
name—Theophrastus   Bombastus   Aureolus   Philippus   von
Hohenheim, or, more simply, Paracelsus (1493/94–1541).

✳ For someone who is as significant to Renaissance science and
medicine as Paracelsus, we actually know very little about him. In
one of his texts called the *Great Surgery,* he says that already as
a youth he had occupied himself with transmutation and that his
father was a most important teacher of the subject. However,
Paracelsus understood transmutation to mean something much
more than turning base metal into gold or silver. In another text
called *Concerning the Nature of Things,* he tells us how the term was
used in the world that he knew best. He writes that "transmutation
is when a thing loses its form and shape and is transformed so
that it no longer displays . . . its initial form and substance, but
rather assumes another form, another substance, another being,
another color, another virtue or property. When a metal becomes
glass or stone, when wood becomes a stone . . . when wood becomes charcoal . . . [or] . . . when cloth becomes paper . . . all of that
is the transmutation of natural things." Changes of this sort
occurred by stages so that the artisan processes of calcination, sublimation, dissolution, putrefaction, distillation, coagulation, and
tincturing could each individually be viewed as producing a kind
of minitransmutation (Paracelsus, 1922–1933; rept. 1996: vol. 11,
p. 349). Whenever you brought something into being that nature
had not entirely fashioned herself, you were doing alchemy.
Paracelsus put it this way. "For nature . . . brings nothing to light
which is completed in itself, rather, human beings have to complete it. This completing is called alchemy. For the alchemist is like
the baker who bakes bread, like the vintner who makes wine, the
weaver who makes cloth. He who brings what grows in nature for

the use of man to that which is ordained by nature, he is an alchemist" (Paracelsus, 1922–1933; rept. 1996: vol. 8, 181). To Paracelsus
it must have seemed that alchemy was an activity that was going on
all the time and was as common as cooking or brewing.

Sometime after leaving home Paracelsus may have come into
contact with a scholarly *magus* named Thrithemius of Sponheim
(1462–1516) whose interpretation of the alchemical writing linked
to Hermes, the *Emerald Tablet,* may have been influential in helping
to fashion Paracelsus's later ideas. Although far from certain, it is
also possible that he received a degree as a medical doctor at Ferrara
in 1515. We are on firmer ground after that date when he gave himself over to a "great wandering" throughout Europe, which, with
only a few interruptions, continued for the rest of his life. Around
1520 we find the first writings bearing his name. One of those texts,
written probably around the same time that he lived in Salzburg
(1524/25) and was much in contact with assayers and miners, was
called the *Archidoxis.* The title is difficult to translate but probably
means something like *Ancient Teaching* or *Deepest Knowledge.* The
important thing, however, is not the title but what Paracelsus had to
say about a new kind of knowledge that could not be learned as an
academic subject. The new sort of learning allowed its students to
discover how to separate the "mysteries of nature" (in other words,
nature's hidden powers and virtues) from material things. Think of
it this way, he said. When an imprisoned man is freed from his
chains, both his body and his individual character or spiritual potential are released. If the man possessed the talents of an artist, he
might then produce a beautiful picture. In a similar fashion, separating the powers within objects from the chains of their bodies
freed nature's hidden talents, and these virtues could accomplish
amazing acts.

This was real knowledge, and power; but it was a kind of knowledge that was nowhere to be found within the standard university
curriculum. There one found only teachers who defrauded philoso-

phy, and who "act as if they were the ones upon whom all belief depends, as if heaven and earth would fall apart without them. Oh, such great foolishness and imposture when they think that they are something that they are not." One needed to give up on this sort of philosophy and instead "seek the mysteries of nature which reveal the end and foundation of all truth" (Paracelsus, 1922–1933; rept. 1996: vol. 3, 95). Discovering the foundation of truth required looking at the world in a new way, and for this reason Paracelsus tied chemistry and philosophy together as the best way to comprehend the real, magical structures of physical reality. The processes of separation (in other words, distillation, calcination, and sublimation) were actually, as Paracelsus saw it, the basic forms of a new type of philosophical knowledge—a chemical philosophy. By means of those processes one could separate the elements, free fifth essences, and also find the healing and perfecting secrets in all of nature. The knowledge of chemical separation was therefore the key to knowledge of both natural philosophy and medicine. Separation led to two types of alchemy. On the one hand, it created what he called medical alchemy. On the other, it led to the alchemy of (metallic) transmutation. By concentrating mainly on the first sort in his writings, Paracelsus strengthened even further the link between alchemy, medicine, and empirical science.

The beginnings of all material things, Paracelsus asserted, were not the elements of Aristotle (earth, air, fire, and water) but the "three principles," or *tria prima*, of Sulphur, Salt, and Mercury. These were as much symbolic categories as rudimentary components of matter. Salt represented an unburnable, nonvolatile ash or earth; Sulphur stood for combustible natures; and Mercury denoted the volatile and metallic constitutions of bodies. Creation of the physical world was itself a process of separation. "The mother and parent of all generation," he proclaimed, "has always been, even from the very beginning, separation." Separation was the first divine act (light separated from darkness), and as such was a miracle that could not be fathomed through human reasoning. Separation from the "great mystery," the stuff of the divine, produced the three principles of Sulphur, Salt, and Mercury. From these were separated the elements and, thereafter, as from maternal wombs, came into being all the earthly, watery, airy, and fiery things of the world (Paracelsus, 1922–1933; rept. 1996: vol. 13, 393ff).

Considering fundamental knowledge to be knowledge of separations is strangely suited to Paracelsus's own life experience in which separation from various communities was a near-constant theme. From Salzburg Paracelsus went to Strassburg, where he appears in the book of citizens in 1526. In 1527 he was called to Basel as a city physician and as a university lecturer, apparently as a result of successful medical treatments after other physicians had given up. Along with lectures in Latin, he offered lectures in German; and that linguistic innovation, in addition to his condemnation of traditional medical authorities (which included burning some of their books), led to sharp confrontations with the Basel community of physicians and brought about his flight from the city in 1528. Shortly thereafter there appeared at Nürnberg two writings dealing with syphilis in which he spoke out against the use of Guajak wood as a medicament, recommending instead a therapy involving mercury. This too proved unsettling to those with vested interests in the older treatment and who therefore had something to lose. Further publication on the mercury treatment was prevented by the medical faculty at the University of Leipzig whose Dean was an intimate friend of the family Fugger, a trading dynasty that possessed a monopoly on the importation of the wood of the South American Guajak tree.

In 1529/30 Paracelsus worked on a book called the *Paragranum*, another hard-to-translate title meaning something like *Beyond the Seed* or *Against the Grain*. The *Paragranum* described the discipline of medicine as resting on four pillars, namely, philosophy, astronomy, alchemy, and the virtue of the physician. Around 1531 he took

up again a book he had begun earlier called *A Work beyond Wonder*
(*Opus Paramirum*), and it is here especially that Paracelsus formu-
lated a new conception of the origin of disease and a new view of
medical treatment.

In contrast to the traditional theory of humors that viewed illness as arising from an imbalance of black bile, yellow bile, phlegm,
and blood, Paracelsus believed that each organ of the body contained an *archeus* (a word with a Greek and Latin derivation imply-
ing a "life power" or "guiding spirit") that acted as an "inner alchemist" and provided for the proper functioning of the organ. All
physicians, he insisted, needed to be able to comprehend the working of this "inner alchemist" and to assist it when necessary because
it was the *archeus* that both maintained health and, sometimes,
caused illness. In the stomach and bowels the inner alchemist transmuted food into nourishment and provided the body with the
foundation for its activity and growth. "God," Paracelsus declared,
"has appointed an alchemist for us to convert the imperfect [which
we consume as food] . . . into something useful to us so that we may
not consume the poison which we take in amongst the things that
are good" (Paracelsus, 1949: 25). When illness occurred, the *archeus*
was usually to blame because, instead of properly separating and
eliminating the poisonous parts of nature, it had allowed something impure to take hold. "Supposing decay has set in in digestion
and the [inner] alchemist fails in his analysis . . . there is thus generated in the place in question a putrefaction, which is poisonous.
For, every putrefaction poisons the site in which it has occurred and
. . . then [that place] becomes a hearth for those diseases which are
subject to it" (p. 30).

Although descriptions like these can be confusing, we should
not to lose sight of what is really going on. Paracelsus is asking one
of the most fundamental questions about the body, namely, how do
its parts "know" what to do? In his view, there must be some guiding principle at work when food is digested and transformed into

blood and nourishment, and, he reasons, illness too must have a lot to do with how well that guiding principle operates. The "inner alchemist" or *archeus* would sometimes fall down on the job and, unless it got help, the body would continue to suffer. This, not humors, is what the Paracelsian physician needed to focus on in treating the body. To do that, the doctor had to learn a lot about the operations of nature—all of nature, because ultimately the operation of the "inner alchemist" was linked to the operations of the world at large and especially to something that had its origin among the stars.

As we have noted, true philosophy, according to Paracelsus, began with a knowledge of the art of separation, the *ars spagyrica*. In his *Book beyond Wonder* he went further and charged a new breed of natural philosopher to understand that "the firmament is within man, the firmament with its great movements of bodily planets and stars . . . Thus what has been spoken of, on the one hand, as pertaining to the firmament, shall, on the other, serve you as an introduction and explanation of the bodily firmament" (Paracelsus, 1949: 36). The knowledge of nature involved an understanding of how each of her parts was designed to "correspond" to specific parts of the human body. Almost a half-century ago, Walter Pagel (1898–1983), in what is still the best book in English written specifically about Paracelsus and his medical philosophy (Pagel, 1958), noted that speculations about analogies and relationships between the world at large (the macrocosm) and the human body (the microcosm) had been around at least since the time of Plato. Paracelsus, however, applied the notion to nature in a new way, viewing the human body as a condensation or synthesis of all the powers of the universe. Within this cosmology, astral emanations pressed on all earthly things (animal, vegetable, and mineral) and gave to them their divinely designated "signatures," in other words, their outward material signs indicating connections to certain parts of the body where they could serve best as medicaments. The network of correspondences and signatures became known through general empirical inquiry and, in a more refined way, by means of analogies established through comparisons to laboratory processes.

If health was good separation, or good chemistry, diseases were a kind of chemistry gone wrong. Just as everything in the macrocosm was born out of the three principles of Sulphur, Salt, and Mercury, diseases of the body were also born into these three universal categories and manifested themselves corporeally as saline (for example, outbreaks of the skin), sulphurous (inflammations or fevers of various sorts), or mercurial (usually diseases associated with a excess of moisture such as phlegm or bodily fluids generally). The important thing to note is that, for Paracelsus, diseases were specific entities with individual characteristics located within particular parts of the body. The way diseases arose in the body, in part due to the shabby work of the "inner alchemist," is one of the most curious and interesting aspects of Paracelsus's medical philosophy; and to comprehend the pathological dynamic involved, you must understand that, according to the system he described, the life of every human being was essentially threefold.

Every person had the mortal life of the physical body. In addition, there was the immortal life that corresponded to the soul and a third life derived from the heavens that corresponded to an "astral body" or "sidereal spirit." This third life was the essential middle link between mind and matter. While not everything in nature possessed a divine soul, all things—plants, animals, minerals, and metals—did possess an astral body, which originated in the stars and which specified that thing's form and function. It was this spirit, or as Paracelsus also refers to it, this *astra,* that penetrated matter and gave life to all growing things, including minerals and metals. It was also this spirit that he viewed as the source of the "the secret forger"—in other words, the "inner alchemist" or *archeus* that, as we have seen, determined and directed the specific functioning of different parts of the body and accounted therefore for the body's overall vitality.

Sometimes the *astra* penetrating the body and determining the functioning of the inner alchemist brought about the generation of something debased and corrupt instead of something wholesome. The source of such spiritually debased generations Paracelsus connected to the fall of Adam. Regardless of the source of the tendency toward impurity, such a generation, manifested in the body as a physical symptom, made a person sick. So illness exhibited the "fruits" of spiritual or astral corruption; and because each disease bore a specific identity as saline, sulphuric, or mercurial and was centered in specific parts of the body that corresponded to parts of the larger world, specific remedies *matching the disease* were required. Remedies cured, in other words, not by counteracting the apparent qualities of illness (hot, cold, wet, dry) with opposing qualities, as in traditional therapies. Medicines supplied spiritual virtues that were drawn from those places in the greater world bearing an affinity or sympathy for the diseased part of the body. As Paracelsus saw it, the medicines were thus able to deliver a kind of specific astral aid where it was desperately needed. Medicaments could be prepared from anything because all things possessed astral spirits or essences that connected them to the macrocosm. However, the most effective remedies were prepared from minerals and metals because these related best to the disease categories manifested as saline, sulphurous, or mercurial. Because illness itself was manifested as a fall from spirituality, healing involved restoring the virtue, or reviving the spiritual vitality, of the inner alchemist, which, in a particular organ or place in the body, had created out of the three principles of Sulphur, Salt, and Mercury something more dead than alive.

The physician, then, needed to be a good observer and to be able to identify substances in nature that corresponded to the pathological "fruit" or symptom of the illness. In this way, like cured like. Yet such substances, in their raw or natural form, might be outright poisons or otherwise noxious to the body. Thus, through processes of alchemical separation in the workshop, the alchemist-physician, as a kind of *archeus* within the outer world, did the job of the inner alchemist, separating the pure from the impure and extracting the spiritual powers from the material dross of an object. He or she thereafter needed to communicate that separated power or virtue to a specific diseased or spiritually debauched part of the body. Walter Pagel called this "medical redemption."

We encounter the localism and specificity of disease in yet another way in one of the earliest theories of illness devised by Paracelsus. This is his doctrine of "tartar." In its final form, tartar could be a calculus, or stone, in the body, or it could refer to any number of bodily changes brought about by the obstructions associated with the heart, lungs, spleen, kidneys, or brain. Whatever the form of tartaric disease, its origin was the result of an incomplete or inadequate separation of the pure and impure parts of substances brought into the body as food or drink. If the *archeus* in the stomach failed to separate completely that which was useful for making blood and bone from that which would be expelled from the body through feces and urine, the impure part remaining was fashioned into a stone or obstruction by what Paracelsus called the "spirit of salt," which he believed to be the ever-present coagulating agent of nature. In this case, treatment involved separation of a different sort, namely, the breaking up of a relationship. In the same way that breaking up a marriage prevents the production of children, Paracelsus said, bringing about the estrangement of tartarous impurities and the "spirit of salt" in the body, through controlled diet and medicines, reduced opportunities for coagulation. The stones and obstructions already produced needed to be dissolved and expelled by means of other medicaments. However, the appropriate instructions for manufacturing these and other remedies,

Paracelsus insisted, could not be found in any library of written texts. One needed "to wander in the library of the whole world, and not just in a part of it, but among all the elements above and below. Such is necessary not just for this kind of illness but for all medical theory . . . Therefore it is required that each person be a cosmographer and a geographer, and that he has tread upon these pages [of the world] with his feet and has seen them with his [own] eyes" (Paracelsus, 1922–1933; rept. 1996: vol. 11, 26–27).

Finally, because the human being was a condensation of the entire universe, Paracelsus thought that an understanding of how the healthy universe of the body worked had to begin with an understanding of how the greater world functioned. The keys to doing this were to be found in philosophy and astronomy, but the pursuit of these two avenues of medical knowledge had very little to do with the way they were usually perceived. Philosophy, for Paracelsus, was not the study of Aristotle, but the comprehension, through experience, of how the forces, virtues, and powers hidden in natural things operated to produce effects of different kinds. Knowledge of astronomy was similarly based in experience of the world, being an understanding of how the powers and celestial virtues linked to the stars and planets affected the functioning of the human body. Philosophy and astronomy as Paracelsus defined them were keys to understanding, through experience, the operations of nature. However, the manipulation of objects so as to produce medical effects required another form of knowledge— namely, an understanding of *chymia,* the processes of preparing useful medicines out of what nature had provided. In this way the physician-alchemist forced things to happen by manipulating the natural world and making use of its hidden powers.

✳ The two terms "Paracelsian" and "hermetic," which I have been using to denote an interplay between the magical and experimental imagination in the Renaissance, are two of the slipperiest terms of the early modern period. Labeling something Paracelsian, for instance, often does little more than to imply some blurry set of ideas linked to the thinking of Paracelsus. The problem is that there is no conformity of opinion about which ideas were original with Paracelsus and which were really expropriated from earlier alchemical and medical authors. Some in the sixteenth and seventeenth centuries did indeed invoke Paracelsus as a forerunner, especially in applying chemical principles to medicine and in turning away from the ancient theory of humors; but it was not necessary to do so because traditions of chemical medicine and even reference to the cosmological three principles of Sulphur, Salt, and Mercury existed as well within other contexts, including medieval alchemy and ancient and Arabic medicine.

Recently, the historian Steven Pumfrey has taken note of a split in historical approaches to matters Paracelsian, and he has divided scholars into two general camps: those who attach the term to a set of core doctrines, spiritual or practical, and those who pick and choose specific parts of Paracelsus's medical-chemical philosophy, usually the practical parts, while ignoring or downplaying what he had to say about magic. What usually goes by unnoticed, however, is that the use of the term "Paracelsian" in the late sixteenth and early seventeenth centuries was frequently meant to convey feelings of disgust and loathing (Pumfrey, 1998).

In the real world of indigenous meaning, the name *Paracelsista* was a label manufactured with hostile intent. It was usually used to condemn those who were viewed as advocating a natural philosophy that subverted trusted Aristotelian wisdom and who had adopted a view of knowledge that embraced magic and the occult. After all, Paracelsus considered that both nature and the human being possessed magical powers. Both were, in this sense, magicians. On the one hand, nature broadcast her secret messages in signs that the magician-physician could listen in on by means of studying disciplines like astronomy, alchemy, medicine, and philosophy. The physician-magus could thereby recognize individual illnesses and create specific cures. In addition, nature also impressed on things a heavenly power that the doctor, by manipulating sympathetic connections, could transfer from one part of nature to another. Furthermore, the *Paracelsista* blended spirit and matter, and mixed thereby the sacred with the profane. Their discoveries and methods did not appear to follow from books or from accumulated experience or structured reasoning. What Paracelsus seemed to rely on most in learning about the world was a kind of divine inspiration; and this, of course, was a process of learning that could not be shared. How do you teach someone to have a revelation, after all? (Hannaway, 1975; Bono, 1995; Pumfrey, 1998).

For Aristotelians (and there were many sorts of these as well), the main issue and horror was the disarray brought to the long-standing symmetry of traditional disciplines by this medley of spiritual, magical, and empirical conviction, and by this most recent attempt to coalesce realms of matter, spirit, and soul into a single subject. To someone like Andreas Libavius, who knew the works of Paracelsus well and who described those who were enthusiastic about them as "neoparacelsians," there was something even diabolical about the whole pursuit. "This very thing," he wrote, "is one of the devil's enterprises, so that he may either abolish or pervert every system of learning, and he himself may rule at his own pleasure" (Libavius, 1613–1615: containing "De Magia Paracelsi ex Crollio," 14).

Indeed, for Libavius the most desperate need at the end of the sixteenth century in regard to the relationship between medicine and chemistry was a clear distinction between which ideas were well established on the basis of reason and experience and which were new, untested, and potentially fraudulent. There needed to be, he thought, some way of separating out the various positions taken by those who had begun to blend together for themselves different parts of ancient authority, medicine, and chemical philosophy. Boundaries were easy to discern in regard to those who followed the teachings of the ancient physicians Hippocrates and Galen. These he called "dogmatists." Only slightly less precise were the lines delineating a group called the "chymiatrists," whom he defined as adding to ancient methods of healing the preparation of chemical medicines by means of alchemy. Some chymiatrists, Libavius observed, had also accommodated alchemy to ancient teachings in medicine by adopting cosmological beliefs in which the microcosm (the human body) reflected the forces and organization of the world at large (the macrocosm). He liked to call these "parabolists," "hermeticists," or even "natural chymiatrists." Paracelsians were birds of yet a different feather. These, according to the world that Libavius knew best, had not only rejected the opinion of Aristotle, Plato, and other ancient authorities, but also had sought novelty in such a way as to embrace the black arts and impious magic, especially the "art of signs" or cabala—in other words, the manipulation of nature using names and words to control spirits (Libavius, 1613–1615: containing "Pro Defensione Syntagmatis Chymici," 1–4)

The thoughts of Paracelsus and Paracelsians did indeed appeal to some who scorned venerated institutions of learning and belief. Healers outside the medical establishment found much to admire in Paracelsus's defense of firsthand experience, as opposed to the authority of ancient physicians, in deciding on how best to learn the medical art. Others found in his writings a source for highly subjective, and therefore institutionally deviant, religious interpretations as well. Many of Paracelsus's theological texts have just recently come to light in collected and published form. In these, as well as in numerous medical writings, he expounded a basic Christian cosmology and expressed many of his ideas by invoking Biblical imagery (Hammond, 1998). Some Protestant groups even found support for their own doctrines, particularly the symbolic nature of the Eucharist, in his writings. Nevertheless, several texts proclaimed views easily denounced as heretical by both Protestants and Catholics. In one such writing, Paracelsus described how God, originally alone in the universe, created from himself a female form (a heavenly woman/wife) in order to produce the second person of the Trinity (Gause, 1991). Such expressions about the nature of God were alarming and even the most nonconformist theologians usually kept well clear of them. There was, however, something else that Paracelsus advocated in these same writings that appealed to religious radicals and enthusiasts—the notion that personal inspiration, not the scriptural authority of organized religious institutions, was the basis of religious insight. The soul's awakening depended on personal revelation, and only then could the words of the Bible, which otherwise remained simply an assembly of "dead letters," be brought to life. Denying traditional religious authority on the basis of a private revelation was just a short step away from denying the authority of secular institutions and the legitimacy of princely power. From the point of view of maintaining political and educational order, Paracelsus and the *Paracelsista* seemed to Libavius, and to many others, to represent the frightening prospect of social chaos and the likelihood of intellectual anarchy (Moran, 1996b; Gilly, 1998).

There is, however, another side to this coin. Sometimes the writings of Paracelsus and others who sought to join alchemy and medicine into a single discipline served to excite a different sort of religious and spiritual reform. In the early seventeenth century, some thinkers began to reconsider an older idea, namely, that beneath the appearance of various and distinct forms of knowledge there existed an underlying unity, a kind of universal knowledge, which, when made part of general education, might lead to the reform of human society. The view was called pansophy and was conceived by a Czech (actually Moravian) minister named Jan Amos Komenský (Comenius) (1592–1670). In England, ideas of a similar sort were also advanced by a German merchant named Samuel Hartlib (ca.

1600–1662) who lived in London and maintained a very large circle of correspondents. Pansophists sought knowledge from all sources and some, especially in the Hartlib circle, looked for clues to an underlying intellectual harmony in alchemy and the spiritual dimension of Paracelsian philosophy. On the one hand, medical alchemy appeared in this setting as a divinely bestowed instrument that offered cures for the diseases that had entered the creation due to the fall of Adam. At the same time, spiritual knowledge acquired through alchemical contemplation could, some believed, ennoble the soul. Once that happened, religious reconciliation and political unity would have a chance to exist and humanity itself, having learned to be more God-like and compassionate, could be transformed (Young, 1998).

✳ While Libavius and others were trying to figure out who was who among philosophers and physicians in Germany, the legacy of Paracelsus and the "art of separation" was undergoing a trial by fire at the University of Paris. In France the medical philosophy of Paracelsus arrived as a result of new texts and old wars. The texts were those of well-respected medical insiders. The wars were wars of religion. Two medical writers especially helped the Paracelsian cause. One was a translator of Galen and a teacher of the famous Renaissance anatomist Andreas Vesalius (1514–1564) named Johannes Guinther of Andernach (1487–1574). The other was a Danish university professor and royal physician called Peter Severinus (1540/2–1602). Both produced books in the same year, 1571, and these helped make Paracelsus at bit more respectable by linking features of Paracelsian medicine to ancient philosophical opinion and by offering more precise descriptions of Paracelsian remedies and ideas.

Severinus's text, the *Idea of Medicine*, sought to place itself within the scholarly tradition of learned medical authority by connecting the doctrines of Paracelsus with those of Hippocrates and

Galen. The result was a more systematic and less eccentric presentation of Paracelsian notions that succeeded so well in altering the complexion of Paracelsus that even the well-known anti-Paracelsian physician and theologian Thomas Erastus (1524–1583) could recommend the text to his readers. In Severinus's hands, Paracelsian natural philosophy became an eclectic mix of traditions linked to well-known attitudes toward nature, especially those expressed by ancient followers of Plato. It was this sort of interpretation that gained the favor of other notable Paracelsians seeking to enhance the reputation of their mentor, such as the professor of medicine at Basel, Theodore Zwinger, and the English physician Thomas Moffett (1553–1604). For his part, Johannes Guinther of Andernach emphasized a more practical approach to Paracelsian therapeutics. As a prominent medical scholar, a translator of Galen, and a professor of medicine at Paris, he prepared an enormous text concerning what he called the old and new medicine that was supportive of chemically prepared remedies. In other places he argued that the chemical principles Sulphur, Salt, and Mercury could be considered to differ only slightly from the ancient elements (earth, air, fire, and water) and judged that chemical procedures did indeed transform poisonous matter into wholesome substances.

Apart from the works of Severinus and Andernach, a further impulse toward the inclusion of chemical medicaments within the practice of medicine in France and elsewhere came about as a result of a famous commentary on the works of the ancient Roman pharmacist Dioscorides, written by the well-known sixteenth-century naturalist Pietro Andrea Mattioli (1500–1577). The text was published in Latin in 1544 and a French translation appeared in 1561. The book included reference to the use of stones, minerals, and metals; and it explained how antimony, which had been described by Paracelsus, could be rendered into an effective purgative. What Mattioli did, in other words, was to situate Paracelsus within an ancient tradition of preparing medicines from minerals and metals.

Where the pharmaceutical tradition of Dioscorides had a place within the medical curriculum, as at the University of Montpellier, instruction in preparing chemical medicines became an altogether acceptable part of medical training. Paris, however, was a different venue with a Galenic faculty harder to please. There a debate over the internal use of mineral-based medicines inspired attacks on suspected followers of Paracelsus. The fear of clandestine support for Paracelsian medical philosophy within the university led to the prompt condemnation of an early advocate of chemical preparations named Roch le Baillif, who had a keen interest in the use of antimony for medical purposes.

Wars are usually good for nothing, but it was the experience of war that also aided the spread of Paracelsus's ideas in France. These were the wars of religion and dynastic ambition that had plagued the kingdom for nearly half a century and that finally brought the Protestant (Huguenot) prince, Henry of Navarre, to the gates of Paris in 1593. There, after accepting Catholicism for the sake of peace, he was acknowledged as the new French king, Henry IV. Returning with Henry were a number of physicians who had been influenced by Paracelsus and who would, in short order, help advance the cause of Paracelsian medical philosophy and chemical medicine. Prominent among this new medical entourage were Jean Ribit (ca. 1571–1609), Theodore Turquet de la Mayerne (1573–1655), and Joseph Duchesne (ca. 1544–1609), who was also called Quercetanus. Although each stirred controversy in his own right, the writings of Duchesne, which expressed a particular variety of Paracelsian thinking, ushered in a renewed period of debate with the medical faculty at the University of Paris.

In his books, Duchesne defended the chemical interpretation of nature, drawing on the universal significance of a microcosm-macrocosm analogy and the underlying creative principles of Sulphur, Salt, and Mercury. The three principles were, in this rendering, just the first of a number of things arranged in threes that linked the realm of the divine with both the heavens and life on earth. They mirrored the union of body, soul, and spirit, and, in a much deeper religious sense, reflected the triune nature of God, the sacred Trinity, from which sprang all existence. However, as much as Duchesne wished to defend Paracelsus in his writings, he also wanted to defend alchemy; and his positions in regard to the truths of alchemy were well known for over a quarter century before the debate about his medical opinions began at Paris. In one debate about the utility of chemical medicines and the origins of metals, Duchesne admitted that the ignorance and faults of some chemists had caused the whole subject of alchemy to fall into disrepute. However, he argued, a few rotten apples should not lead to condemnation of an entire art in which God had revealed so many secrets of nature and so many preparations of herbs, animals, and minerals. There was also no good reason to deny the possibility of transmutation.

Alchemy, in Duchesne's view, was a profound and hidden part of *physica* (medicine), which promised to those who understood it an intimate knowledge of preparing remedies for the protection of human life. The metaphor of transmutation became for him a powerful image by which to view the therapeutic aim of the chemical physician. No one should think, he proclaimed, that when he used terms like the "universal balsamic medicine," the "fifth essence," or the "celestial stone of the philosophers," he was referring to something that would transmute metals. What he was after was transmutation all right, but a transmutation of a different, internal sort. "But knowe rather," Duchesne is made to say in an English translation, "that in man (which is a little world) there lie hidden mines of imperfect metals, from whence so many diseases grow, [and] which by a good faithful and skillful Physician must be brought to Gold and Silver, that is to say, unto perfect purification by the virtue of so excellent a medicine" (Duchesne, 1605: G4v; Debus, 1991; Kahn, 2001).

Such things were troublesome at Paris; and even though

Duchesne insisted that chemists valued traditional medicinal ingredients and found much to praise in Galen and Hippocrates, the Faculty of Medicine, through its spokesperson Jean Riolan (1539–1606), wasted no time in responding to and condemning his ideas. Riolan was accustomed to debate and had recently emerged from another controversy concerning certain kinds of diseases labeled occult that had been championed by the so-called French Hippocrates, Jean Fernel (1497–1558). For Fernel, who explained his views in a book published at Venice in 1550 called *Two Books Concerning the Hidden Causes of Things,* the true source for the powers of the living body was to be found not in the actions of the elements or humors but in a "total form" or "vital heat" that was divine in origin. The vital heat was also a celestial heat and made use of a heavenly spirit to affect the functioning of the body. Diseases that weakened the functions attributed to the divine or innate heat and that were thus not a consequence of the body's imbalance of qualities (hot, cold, wet, and dry) were deemed diseases of the "total substance" for which appropriate remedies hidden in nature needed to be found. Fernel's books blended practical medicine with speculative natural philosophy, yet Fernel himself was always able to maintain a position of respect among Paris doctors. What he considered to be diseases of an occult nature (contagious or pestilential diseases) others, including Riolan, assigned to the actions of various corruptions or poisons—none of them, however, seen as descending from the heavens.

Duchesne suffered attacks from different quarters at Paris, but he also had his defenders, and one of the most prominent was Theodore Mayerne, another court physician (Cook, 1986: 95ff). Mayerne argued that medical knowledge progressed through experience and, while truth was open to all, it had not yet been seized by anyone in its entirety—not even by Parisian professors of medicine. For the true physician, experience always reigned over method; and it was for this reason that, he writes, he had committed himself to extensive travel and had discovered thereby the art of alchemy, the nurturing mother of all experience. The mayhem seemingly caused by Paracelsus and "Paracelsian" ideas at Paris Mayerne suspected was really caused by something else. The real question being asked in the Parisian debate was this: Should alchemy be accepted as an independent discipline, which, because of its powers of understanding the operations of nature and the body, was not merely a part of medicine but *reigned over* medicine and provided medicine with a new, chemical, rationality? Put another way, the question was even more disturbing. Should there be a faculty of medicine at Paris, or should there be instead a faculty of alchemy? Our friend Andreas Libavius also knew that this was what the fuss at Paris was really all about. The official censure of Duchesne and other Paracelsian physicians by the Parisian faculty, he wrote, was not about Duchesne at all. The real problem was whether alchemy provided a better overall understanding of the workings of the body and better ways to maintain health than other, more ancient forms of medical wisdom. The censure at Paris had been pronounced, Libavius proclaimed, "not on account of Quercetanus ([Duchesne] but because of alchemy" (Libavius, 1606: containing "Commentariorum Alchymiae . . . Pars Prima," 1ff).

✳ Sometimes, then, apart from conceiving of illness and its treatment in new ways, following in the footsteps of Paracelsus meant to be of the opinion that the best way to know the body and to understand its functioning was by means of chemistry. This aspect of what later became known as "iatrochemistry" developed in the seventeenth century within the writings of numerous authors. A few deserve special attention; however, one in particular should be discussed at this point. This is the Brussels-born physician and chemist Jean Baptiste van Helmont (1579–1644).

Although van Helmont followed in the tradition of Paracelsus, and seems as well to have been influenced by Peter Severinus, there were also marked differences between his views and Paracelsus's beliefs. In particular, van Helmont rejected the analogies linking the macrocosm with the microcosm and refused to think of the Paracelsian first principles as preexistent in material substances. Sulphur, Salt, and Mercury were instead generated in substances by the application of heat, he concluded. Moreover, while still accepting the existence of sympathetic attractions in nature, van Helmont believed these to occur naturally and not as a result of supernatural forces. This last view brought him, in 1621, into an already-raging controversy concerning the so-called Paracelsian weapon salve (an ointment that supposedly cured wounds through magical sympathies after being applied not to the wound itself but to the weapon that had caused it). Some readers may recall that Umberto Eco used a similar sympathetic relationship based in the thoughts of Paracelsus, involving a sword and a wounded dog, in his novel *The Island of the Day Before* (1995). As the book explains, finding one's longitude at sea was tough business in the Renaissance unless one could compare the time of day aboard ship to the time of day back home. In Eco's story the difference in time zone was determined by means of the magical natural sympathy that existed between a wound and the thing that caused it. A wounded dog was taken aboard ship and the wound kept open throughout the voyage. At certain prescribed times the sword that had caused the wound and that was kept back at port was plunged into a fire. The sympathetic connection between sword and wound caused the dog to feel pain and to howl. Simple computation thereafter led to a determination of longitude. Perfect magical reasoning, although hard on the dog.

Eco's story is, of course, a fantasy; but the weapon salve controversy was not. In his own contribution, van Helmont concluded that a certain "magnetic" sympathy existed not between the weapon and the wound, but between the wound and the blood left on the weapon. Something similar to this type of magnetic sympathy, he felt, also accounted for the effects of sacred relics. These views, when linked to a Paracelsian philosophy of nature in which sidereal spirits were believed to impinge on everything in the terrestrial world, proved to be his undoing. Van Helmont was condemned by both the faculties of medicine and theology at Louvain and twenty-seven of his "propositions" were found to be heretical by the Spanish Inquisition. Thereafter he was imprisoned and, later, sentenced to house arrest. The fact that church proceedings against him formally ended only two years before his death prevented the publication of most of van Helmont's chemical-medical ideas during his lifetime. His collected works came to light only after his death, edited and published by his son, Franciscus Mercurius, under the title *The Origin of Medicine* (1648).

Much of van Helmont's medical philosophy was concerned with the activity of vital spirit in nature. All things in nature, he believed, arose from spiritual seeds planted into the medium of elementary water. The seed also possessed the life force of all animals, vegetables, and minerals. By means of a ferment, which van Helmont described variously as the beginning of all things and as that which determined the form, function, and direction of every existing thing, the seed transformed water into an individual being. In respect to disease, he thought of each illness as also a specific thing produced from a particular "seed" that had been fertilized by an enfeebled vital principle.

To find the invisible seeds of bodies, van Helmont attempted to explore chemically the smoke arising from combusted solids and fluids. It was this "specific smoke" (in other words, that which differed from air and contained the essence of its former material substance) that he termed "gas." The term as we encounter it today has lost most of the meaning that van Helmont gave it. As with the idea of spiritual seeds, or *semina,* his ideas grew from a context teaming with thoughts and formulations that merged divine action with physical existence. "Gas" was another illustration of this connection, linked to the assumption that nothing was entirely inert in nature, and that in every one of her parts there could be found a spiritual life giving or activating presence. This sort of natural philosophy is often referred to as vitalism; and another term coined by van Helmont, called *blas,* even more clearly connects to such a vitalist conception of nature (Pagel, 1982). *Blas* represented a universal motive power, present everywhere in nature (Debus, 1977; rept. 2002: 295–339). Some of it was derived, he thought, from the stars, but some of it also was innate in living things. In human beings the *blas* was both internal and external. Human beings possessed double *blas.* They received life from the heavens but also exhibited vitality according to the use of free will: "One [*blas*] to be sure, that exists by a natural motion, the other truly [a thing] of the will, because by means of an internal willing it exists as the motor to itself . . . without [requiring] the *blas* of the heavens" (van Helmont, 1667: 112). The motive powers of the universe gave rise to the possibility of life and action. Human beings, however, were free agents. Although moved physically by the powers of nature, through generation, growth, and inevitable death, they maintained an inner motive power, a will, that separated them from the rest of creation.

While careful to maintain a distance between his own ideas and Paracelsus's explanations of disease, van Helmont nevertheless shared with many of Paracelsus's followers the belief that the key to understanding nature was to be found in chemistry. "I praise a generous God who called me to the art of the fire . . . For, more than all the other sciences, *chymia* prepares the intellect for penetrating to the hidden parts of nature, and thus penetrates to the furthest depths of objective truth" (p. 286). Hands-on experience with laboratory procedures led van Helmont to give a good deal of attention to determining the weights of substances in chemical reactions. Against Aristotle, and on the basis of observations of a burning candle surrounded by a glass container resting in water, he argued that air could be diminished or contracted, thus making possible the existence of a vacuum in nature. He also advanced techniques for various chemical preparations, especially chemical medicines involving mercury, and advocated a view of matter as made up of tiny particles or corpuscles. Following suggestions found in the writings of Paracelsus and Duchesne, he determined that acid was the digestive agent of the stomach and noted further that alkali acting on acid exhibited neutralizing effects. These observations would have enormous theoretical consequences in the years to follow, and we will have more to say about them later. More important at this point, however, is to note that the "art of the fire" revealed to van Helmont other, more deeply hidden truths as well. Thus, in his major text he also gave attention to the transmutation of metals, to techniques for separating the pure from the impure parts of nature, and, of special significance, to a substance, called the *liquor alkahest,* which he accepted as one of the greatest secrets of Paracelsus and which he referred to as an incorruptible dissolving water that could reduce any body into its first matter (p. 481).

Almost everyone knows of van Helmont's famous tree experiment, in which he compared the weight of water given to a growing tree against the weight of the tree itself; but what sometimes goes by unnoticed is that such use of quantitative evidence and experimental design existed quite nicely in van Helmont's natural philosophy next to ideas like *blas,* the Paracelsian weapon salve, and the marvelous *alkahest.* Both aspects of van Helmont's approach to nature, the quantitative experiment and the devotion to vitalism, became part of his legacy and influenced scholars and lay readers all over Europe. By 1707 twelve editions of the *Origin of Medicine* had appeared in five languages and had inspired others to think of the functions of the body and the origins of disease as analogous to chemical operations observed in the laboratory. But something very interesting occasionally happened to van Helmont's ideas when they entered the vernacular neighborhood of the later seventeenth century. Vitalist references like *archeus* and *semina* were still there;

but, as in the treatment of van Helmont's ideas by the Oxford medical doctor Walter Charleton (1619–1707), they kept different company, consorting with elements of what came to be called the mechanical philosophy (Clericuzio, 1993: 306ff)—a view of the physical world in which sidereal spirits and sympathetic correspondences no longer had a place and in which all things were to be explained solely in terms of matter and motion.

✳ It is tempting to assume that the mechanical philosophy and the new experimental science of the seventeenth century were so antithetical to vitalist and magical beliefs that no representative of the new science would be able to tolerate Paracelsian ideas for long in any sort of public forum. That may certainly have been the case for many advocates of the mechanical philosophy. But when such a view is expressed without exception, think about a book, published in 1691 by a certain Hugh Greg, an amanuensis (secretary) to the famous experimental chemist Robert Boyle, called *Curiosities in Chymistry: Being New Experiments and Observations Concerning the Principles of Natural Bodies.* The book, a popular potpourri of chemical, anatomical, and philosophical opinion, sold lots of copies and was reprinted twice more in the 1690s. Many people read the book; and when they did they found in it a remarkable juxtaposition and combination of Paracelsian theory, contemporary anatomy, corpuscular philosophy (in other words, the view that physical matter was made up of tiny particles), and analytic chemistry—all sharing the same textual space without doing violence to one another, even if not depicting a consistently unified vision of nature.

According to the author, who based many of his ideas on the views of van Helmont, water is the first material principle of all mixed bodies. "Seeds," on the other hand, which contain the blueprints for what a certain thing will become, determine the specific form and purpose of every body. These two components of things (water and "seeds") are united together by means of "acid ferments." So, water is coagulated into a plant by the ferment of a vegetable seed; into metal, stone, and so on by that of a mineral seed; and into flesh, bones, and so on by the ferment of an animal seed. "For in all mixt bodies there are certain . . . particles wherein the seeds or Ideas of natural things do reside, and which, do follow the draft [design] of these Ideas" in giving forms to that which is vegetable, mineral, or animal (Greg, 1691: 29).

Something similar could have been written by any number of Paracelsian authors. The notion of seminal "ideas" giving form and function to material stuff has, apart from van Helmont's discussion of them, much in common with the thoughts of the earlier Danish Paracelsian, Peter Severinus, who also expressed a notion of "seeds" containing the "idea" of the thing they were to become. The big difference in Greg's description of "ideas," however, is that these "ideas" are actually particles, or at least are contained in particles. They play their part not in a Renaissance setting of magical relationships, but in the mechanical and rational order of seventeenth-century experimental science, fitting in among the most recent anatomical discoveries, including William Harvey's discovery of the circulation of the blood. "The chief mover (under God) of all natural bodies," says Greg, "that coagulates elementary water into all sorts of bodies, according to the various ideas of those seeds . . . is a certain subtil spirit of an igneous nature, diffused through the whole visible world." However, he adds, "by spirit here is not meant an immaterial substance but a body consisting of very minute and very active particles, peculiarly fitted for motion." This is a quick change of huge consequence. "Spirit" has turned into matter—a very subtle matter, but matter nevertheless. It was the material particles of bodies that, at least for Greg, contained the information of how specific parts should take shape and should function thereafter. With such an understanding, the generation of all things, even the riddle of human reproduction, stood open to possible explanations not conceived of before.

"Every part of a woman's body," says Greg, "has its own Idea residing in it," and it is a particle of this "Idea" (in other words, this plan or organizing data) that is communicated to the blood as it circulates through that individual part. "The blood [then] carries all these ideas to the testes [ovaries] where they are gathered together, disposed into the same order that the parts they come from have . . . and so [are] united into one entire Idea, which is enclosed within the tunicles of the egg." Thus "if it were possible for us to contemplate the Idea with our bodily eyes as well as we can do with our intellectual [eyes] we might discern . . . all the parts of the body as [if it were] an exact model, or an entire woman" (pp. 63–64).

Like the female idea-particles gathered in the egg, masculine particles in the reproductive seed contained their own ideas, but these ideas are, says the text, "confused." They are confused because not being enclosed in eggs, but rather being contained in the testicles in liquid form, "they fluctuate and cannot retain any certain order." "Hence it is, that as the feminine seed alone can never be fruitful . . . so neither can the masculine seed alone ever produce a foetus, till its confused ideas be reduced into due order by conjunction with the feminine" (pp. 64–65).

At this point, notions derived from Paracelsus and van Helmont become interwoven with the most recent discoveries of observational anatomy. The masculine seed being injected into the newly identified *ovarium* causes one or more eggs to be impregnated and thrust into the extremity of the Fallopian tube *(Tubus Fallopianus)*, which conveys the egg(s) to the womb. Only by means of the heat of the womb, however, are the seminal ideas in the egg excited into motion. The process of gaining physical form occurs thereafter through coagulation "by which means the ideas, that were utterly insensible before, do quickly acquire a visible bulk" (pp. 65–66). The sex of the foetus is determined by which ideas in particle form, those of the father or of the mother, are greater in number when mixed together.

That a woman's imagination during conception could affect the formation of the foetus was a notion as old as antiquity and one dear to Paracelsus and other medical authorities throughout the Scientific Revolution. A French contemporary of Paraclesus, the well-known surgeon Ambroise Paré (ca. 1510–1590), noted in his famous text *On Monsters and Marvels* (1573) that the images of animals or of different kinds of food might appear on the body of the newborn due to "the force of the imagination being joined with the conformational power, the softness of the embryo, ready like soft wax to receive any form." Paré further instructed that "women—at the hour of conception and when the child is not yet formed—not be forced to look at or to imagine monstrous things." Failure in this regard could result in hideous offspring such as a child with the face of a frog, which, he reported, was produced by a woman who conceived while holding a frog in her hand as a remedy for fever (Paré, 1573; rept. 1982: 38–42, 54). Our text takes a different approach, but one that is no less fascinating. It explains the same phenomenon in terms of a mechanical, and allegedly completely rational, view of nature. Because the sight of an object is first painted on the retina by rays of light reflected from an object, the same scene, by means of subtle particles, might be conveyed from the brain to the testes (ovaries) and there impressed on the seed. "For if the [particulate] spirits of the optick nerves transmit this idea from the eyes to the brain and there imprint it; why may not . . . the *par vagum* [one of the branches of cranial nerves] transmit the same idea from the brain (through certain little branches that reach) to the Testes, and there communicate it to the seed?" (pp. 70–71).

Hugh Greg is not one of the stars of the Scientific Revolution. Few people have, in fact, ever heard of him. His book provides nevertheless a useful example of what could happen to learned discussions when they entered the popular mind. According to Greg, the best and latest anatomical observations served to support a mechanical explanation of the operations of the body. At the same

time, the discovery of the circulation of the blood and a corpuscular view of nature offered a new basis from which to argue for the
influence and power of imagination in the process of reproduction.
As much as a mechanical outlook on nature liked to keep itself
apart from principles of vitalism in official circles, Greg's book tells
us that the separation was not necessarily absolute. Some writers
did indeed detach themselves from reigning philosophical norms
and chose instead attitudes of creative accommodation in interpreting the processes of nature. As we will see later, even some of
the most enthusiastic mechanists, such as those who sought to
explain natural phenomena solely in terms of matter and motion,
needed to think a little outside the box when faced with explaining the artfulness or ingenuity of nature and with comprehending
the workings of the parts of that animate machine called the human body.

CHAPTER FOUR

SITES OF LEARNING AND
THE LANGUAGE OF CHEMISTRY

It's tempting to want to determine precisely the points at which the historical thing we call the Scientific Revolution happened. In terms of the relation of alchemy and chemistry to the construction of natural knowledge during the Scientific Revolution, what "happened" was not an event, or series of events, at all; it was not a particular thing or idea that was there at one point and not before. Instead, it was a subjective reevaluation of experiences that had been around for a very long time. The same is true in other contexts, of course. Both Galen and William Harvey could cut open an animal's body and find the heart, but Harvey saw the heart as a pump and described the veins and arteries as a mechanical structure for the transport of the blood. Harvey reevaluated the ancient experience of dissection, and living in a world of water pumps and other machines helped him find the appropriate metaphors to do so. Although not our immediate purpose, we could go further and interpret the Scientific Revolution itself as a process of rethinking older experience. In that case, far more important than isolating any static event or specific book that supposedly changed everybody's view of the world would be to understand how a reassessment of various kinds of experience (some of it related to alchemy) could have resulted in new perspectives about how the world worked. Beyond that, it would also be valuable to know precisely how such a reevaluation came to be communicated among individuals, each personally prepared to receive it.

Shifting the perspective of experience does not happen overnight. Neither does it happen usually without a fight. Take, for instance, the experience of our generally recognized paragon of good chemistry, Andreas Libavius. You may recall that he wrote a book called *Alchemy* at the close of the sixteenth century, which, despite the title, some have called the first real textbook of chemistry. This is not the only book that Libavius wrote, however. Among many others, he also composed what looks from the title to be nothing more than a collection of letters. Yet, despite the plain wrapper, there is something very important going on inside the book's covers telling us that a shift in evaluating alchemical and chemical experience had begun.

In the very first letter of the collection, Libavius lets us know just what he thinks of chemists and chemistry. What he has to say may seem surprising. He first asks, "What is more abject than a chemist?" and then proceeds to define the chemist as "the enemy of nature." The chemist, in his view, was a horrible, morally corrupt person and there did not seem to be any single term awful enough that could be used to describe him. Therefore, whatever accusations could be found, he advised, should be "all piled together and hurled at the professor of chemistry." The more excellent one was as a philosopher, the more one needed to separate oneself from the "cohort of chemists." You had to be really "insane" and altogether "studious of vanities" to be attracted to chemistry. Chemistry was "a bilge-flood and chaos of impurity and human dregs." In no way could its practitioners be granted a place among philosophers. "Could you even stand to walk . . . with such a fellow?" Libavius asks. "Would this man even be worthy of life?" (Libavius, 1595–1599: book 1, 9–16). This is certainly not the perspective of chemistry that we share today. So, what is going on? How could someone fashion such a view, and in what circumstances does the experience of chemistry get reassessed?

We have to read the second letter in the book to discover what Libavius is getting at. There he says that he has discussed chemistry in such a way that one might think that he was attacking the subject altogether. However, the real meaning of this particular art was not what many people thought it was. "Chemistry," at least in Libavius's opinion, had become a subject almost entirely in the hands of frauds and impostors, and these had horribly changed its likeness to medieval alchemy. Libavius pulled no punches. Just as the essence of a woman, he argued, was not found in being a prostitute, so the essence of chemistry could not be determined by those whose only talent was in deceiving the public when they promised to produce quintessences of gold and other remedies in the manner of Paracelsus (book 1, pp. 19–20). The term *chymia* (chemistry) had fallen into disrepute. It had come to be a cover name for "novelties"; and from the point of view of this celebrity of modern chemistry, nothing was worse than to be counted among those called the "moderns" (*recentiores*). The moderns, he said, were a seedy group whose members agreed with no one and who condemned all the writings, assertions, and deeds of the past. The worst of the lot was Paracelsus. He had claimed for himself the monarchy of chemistry; and in so doing, he had opened up a way for accidental discoveries, or secrets, to count as real knowledge. Hence, it had come to pass, Libavius continues, that hardly anyone agreed with anyone else, and each person wanted to seem to have brought forth something new *(aliquid novi)*, the knowledge and art of which he laid claim to only for himself (Libavius, 1613–1615, containing "De Alchymia Pharmaceutica," 127). To give chemistry its proper place among the sciences, one could not look at it in this way. One had to take a different view.

Real chemistry was not a novelty or secret. It was not "modern." It was an art rooted in the books of ancient philosophers and in the accumulated experience of artisans over hundreds of years. *Chymia* (concerned largely with making medicines) was, in Libavius's opinion, really a branch of alchemy and thus appropriate to its study were the texts of a large number of medieval and contemporary alchemical writers. However, for all that, he concluded that the true art of *chymia*, and the rules by which it could be taught, was really nowhere yet to be found. Hence, Libavius says, if anything was to be written about chemistry it had to be reduced almost to ABCs, and before learning any concept, one had to learn what sort of thing "chemistry" was. (Libavius, 1595–1599: book 1, preface, to the reader). Libavius's plan was to write letters to friends and acquaintances about chemical procedures and terminology and to formulate new meanings on the basis of collective experience. The result would not only make public what Paracelsians liked to keep secret but would, once and for all, define the language of chemistry. Libavius's letters, published in three parts between 1595 and 1599, really defined the state of the art; and if anything can be regarded as essential to the origin of chemistry as an academic discipline, it is not his *Alchemy* but this volume with the very bland title *Book of Chemical Subjects* (1595–1599).

Chymia had been enslaved and prostituted by Paracelsian physicians. But how to deliver her from the intellectual brothel where she was now imprisoned, and where could she be safely kept thereafter? The struggle that Libavius and others took part in over the possession of *chymia* was as much a contest of competing philosophies and methods as it was a quarrel about which institutional authorities would legitimize their use. For Libavius and others who attempted to break the hold of Paracelsian magic on chemistry, some places, or sites of knowledge, were preferable to others. One was esteemed best of all, and one was necessary to avoid at all costs.

✳ Paracelsian physicians often relied on court positions to establish both the social and intellectual credibility for their theories and practices. One of the first publishers of Paracelsus's works, Adam von Bodenstein, was court physician to the German prince Ottheinrich. Later, the chemical investigations of one of the best-known Paracelsians in Germany, Oswald Croll (ca. 1560–1608), led to the creation of what many people would consider to be the most important compilation of Paracelsian remedies, a book called *Royal Chemistry or Basilica Chymica* (1609). The book took shape with the financial support of the Calvinist prince Christian I of Anhalt-Bernberg. Croll, who was appointed court physician by the prince and who acknowledged his debt to Christian in the preface of the book, was also one of the court's most important agents and one of its chief negotiators. At many small German courts as well as at the mightier courts of the Holy Roman Emperors in Prague and those of the French kings Henry IV and Louis XIII, Paracelsian physicians had come into real prominence. The appearance of Paracelsian "chemistry" at the Spanish court of Philip II followed the creation of court-sponsored distillation laboratories and the appointment of physicians and apothecaries to court positions who were sympathetic to Paracelsus's teachings (Bueno and Pérez, 2001). Court patronage of artisan alchemy was one thing, but support of Paracelsian chemistry, with its secret remedies and magical associations, was another; this is where Libavius drew the line. Responding to one of his many adversaries in 1594, he asked, "If you are real doctors and do not flee from the light [of truth,] why do you not teach in the academies? You are occupied at courts where I believe you accomplish more by flattery than by speaking the truth, and advance more by begging than by curing" (Libavius, 1594: 729).

The significance of courts and courtly cultures in the promotion and fashioning of innovative ideas in the sixteenth and seventeenth centuries, especially ideas related to the history of science, has been well established. And yet, from the point of view of reevaluating past experience and establishing the essential meaning of chemistry, no site was worse. Chemistry would not gain its intellectual legitimacy through courtly ties. To forge the values of real chemical knowledge, there could be no noble title to truth. Chemistry, once delivered from the cultures that had debased it, needed to be dressed in academic gowns and accepted within the university curriculum.

This was an enormous task. Along with calling for openness and utility in chemistry, Libavius and others needed to amplify the moral and political power of universities and academies in defining the norms and values appropriate to a particular way of gaining knowledge and in determining what was legitimate to chemistry and what was not. Yet, chemistry's entry into the university was clearly impossible as long as it remained linked to magic and was separated from traditional Aristotelian philosophy. Libavius wondered, How could chemistry, in its present shape, ever be deemed worthy of a place among liberal studies? To make the subject fit the site, not only did chemistry need to become more philosophical and open to didactic method, but the academy itself needed to redefine what was appropriate to its curriculum as well.

What, then, was chemistry to be? Basing his views on what had already been written about alchemy by medieval authors, Libavius considered that the job of the chemist was to resolve or break apart by art the things that had been mixed together by nature. Thereafter, the task was to purify them according to what had come to be known through experience and observation. Furthermore, the purpose of chemistry was to exalt those things that were already pure in themselves but that had not yet acquired powers to suit a certain purpose (Libavius, 1595–1599: book 2, 6–12). Most important, chemistry was to stand alone as an independent subject, content in itself and possessed of its own intellectual domain. Libavius adds that when natural philosophers disputed about motion, infinite space, or the existence of a vacuum, their discussions resulted in advancing the knowledge of individual things no further than could be achieved by someone of mediocre learning. "The chemist," on the other hand, "has investigated sympathies and antipathies, causes, effects and the rest of nature one [part] at a time, and thus does not know them indefinitely and vaguely, but definitely and certainly . . . If therefore you thought previously that chemistry was nothing more than a handmaiden to medicine, correct your opinion, and consider this to be one of the most worthy arts" (book 2, pp. 39–40).

Libavius's central rule, or motto, for determining what belonged to chemistry and what did not followed from Aristotle. The guideline would become one of the most important statements for the further development of chemistry as a legitimate academic subject. The rule was simply this: "Nothing is to be received into chemistry which is not of chemistry" (book 1, p. 119). That looks so simple as to appear silly. The implications, however, were enormous because, according to this view, magic, celestial influences, and divine revelation were out of bounds. If chemistry was about the mixtures of the material world, then what was appropriate to the subject of chemistry and what should count as chemical knowledge had to be found entirely in the physical stuff of the earth.

As the practice of chemistry became, in this way, more philosophical, philosophy needed to open itself up to more than just strict contemplation. Thus, Libavius used several letters in his books of correspondence to describe a philosophically based chemistry in which manual operations actually ennobled philosophical study. To one correspondent, he quoted the ancient Roman writer Cicero, saying that all praise for virtue consists in action. So, he continues, "Is our chemist better off observing rather than acting?

If you think this, then come, let us occupy ourselves day and night in philosophers' books, and let us gladden our souls in the contemplation of chemical operations and [different sorts of chemical] species. We will leave manual tasks and tools to the furnace shop. For the chemist should have his mind worn down by knowledge, not his hands . . . [but] what aid [then] is he to humanity who only delights himself by thinking up images in his head and contriving things which cannot be put to use? . . . So, although we think that the pleasure of chemical contemplation is special, yet this art also demands working [with the hands], both to strengthen theory (since theory can be deceived by the vanity of opinions) and also that something beneficial to all might result" (book 3, pp. 10–15).

To dress chemistry in academic garb required a community effort and, at least in 1595, the best person to lead that community seemed to be a professor at the German University of Jena named Zacharias Brendel (1553–1626). Brendel had erudition, eloquence, experience, and method. Libavius believed that, just as astronomy and logic had been reformed during his lifetime, so too could Brendel provide the fundamental principles and general precepts for constructing a logical method for chemistry. "Tycho Brahe [the astronomer] is said to be the restorer of astronomy," he wrote. "Why not win praise for yourself from chemistry? . . . We shall piously set up a statue in honor of chemistry for you if we owe the perfection of the art to your industry" (book 1, pp. 115–16).

Chemistry of a sort did come to the university at Jena. Both Brendel and his son (also called Zacharias, 1592–1638) taught how to prepare chemical medicines there, and they were followed in that sort of teaching by Werner Rolfinck (1599–1673), who acquired the more specific title of "director of chemical exercises." But limiting instruction in chemistry to medicine was not really what Libavius had in mind. And while it is true that, as the historian of chemistry Maurice Crosland and others have argued, the real status of chemistry as an independent discipline within the university had to wait for the eighteenth and even the nineteenth century to be secure (Crosland, 1996), attempts were made much earlier to find a place for chemical knowledge within academic walls. For Libavius, the academic institutionalization of a non-Paracelsian kind of chemistry was made all the more urgent because of something shocking that happened in the year 1609—something that he had until then always thought impossible.

✳ In that year, what Libavius and others called chymiatry (chemical medicine) gained a foothold within the medical curriculum at a university in the German town of Marburg. That was not the astonishing thing, however. Marburg was not alone in offering such practical courses in making medicaments. Instruction in the preparation of chemical medicines could already be found elsewhere, both in and outside Germany, as part of medical education. As we have seen, a pharmaceutical tradition based in the works of Dioscorides allowed a student to learn how to concoct chemical medicines at the University of Montpellier. In Spain, the professor of surgery Llorenç Coçar (ca. 1540–1592) was named to a newly created position designated specifically for teaching the "secret remedies of diseases" at the University of Valencia. The course that he created centered on the preparation and administration of chemical medicines, but it was offered only once before Coçar's disappearance or death in 1592. Back in Germany, the well-known academic physician Daniel Sennert (1572–1637) introduced chemical instruction into the medical curriculum at the University of Wittenberg sometime after joining the faculty in 1602. Sennert's writings centered, in part, on reconciling Aristotle, Galen, and Paracelsus. At the same time, however, his own medical philosophy remained firmly rooted in ancient traditions.

Aristotle may still have been on stage at Marburg, but he was not in the spotlight as far as teaching chymiatry was concerned. There, the man appointed to be "public professor" of the new discipline of *chymiatria,* Johannes Hartmann (1568–1631), was a favorite of the German prince Moritz of Hesse-Kassel; and Moritz, everyone knew, was a well-known patron of hermetic and Paracelsian ideas. Hartmann also embraced divine mysteries and used a magical symbol as his personal letter seal. This was the shocking part of Hartmann's appointment. A Paracelsian was going to teach *chymia* within the university. Libavius probably had Hartmann in mind when he wrote that chemical studies had progressed to the point at which a youth sent into the academy needed no longer to worry about having to labor over the decrees of philosophy. Instead he could inquire into the novelties of those who otherwise hid their art from view. While knowing nothing of the alchemy of medieval writers, and being ignorant of logic and the works of Aristotle, students were instead trained to admire diversions born scarcely yesterday from furnace smoke (Libavius, 1613–1615: containing "De Alchymia Pharmaceutica," 127–128).

Regardless of the ridicule, Hartmann's instruction in the art of chymiatry (chemical medicine) was one of the earliest examples of laboratory-based chemical teaching within a university curriculum. We know a lot about Hartmann's classes because a description of the rules for his laboratory (students were required to leave their swords at the door) and an account of the procedures taught to his students still exist for two semesters in 1615 and 1616. From this account, it is clear that Hartmann relied a great deal on recipes adopted from one of the most popular Paracelsian formularies of the early seventeenth century, the *Royal Chemistry* (1609) of Oswald Croll (ca. 1560–1609).

The first part of Croll's book has a lot to say about magical relationships in nature; and students, if they read it, probably found it difficult to understand. The second part, however, is dedicated to the practice of chymiatry and it is there that Croll gave precise procedural directions for the preparation of his many medicines, along with directions for their use and dosage. While part of the book, then, focused on correspondences and sympathies, it would not have been too difficult for Hartmann and his students to ignore or detach the magical bits and to concentrate on the more practical sections. The book, as Paracelsian as it was, thus took shape at Marburg, and probably elsewhere too, as a text suitable for teaching.

✳ If you were a professor or student of the faculty of medicine at the University of Marburg in the early seventeenth century, odds are that on April 4, 1609, you had made plans to attend a public oration to be delivered by the new "public professor of chymiatry," Johannes Hartmann. If you had been tempted to stay home that day, two things would probably have made you think twice. First, this was an appointment of the university's protector, the prince of Hesse, Landgrave Moritz. Second, the title of the lecture was curious. In it Hartmann referred to himself not as a Galenic doctor but as a "philosopher, or skilled natural physician" (*philosophus, sive naturae consultus medicus*) The title was revealing because, as you would have soon discovered, what this new member of the school of medicine had in mind was the training of a new sort of medical doctor, someone who would have become experienced in the laboratory, would have known how to make useful medicines, and would have known which parts of ancient medicine could be combined with alchemy and the ideas of Paracelsus to produce healing remedies. Paracelsian medical philosophy, a thing that Libavius and others had long thought of as close to philosophical insurrection, had found a way to slip into the university.

Hartmann promised that his students would comprehend the intimate harmonies of the universe and understand the analogous relationship between man (the microcosm) and the macrocosm. The library, it seemed, was not going to be the only place where this new kind of medical student could be found. Hartmann proclaimed that the "natural physician" had to dwell in the whole world, fly over its seas, and burst through the ramparts of the heavens. Then one would know "that the heavens, stars, and all airy, aquatic, and terrestrial things are lodged in man"; that "lightening, thunder, hail, rain, heat, cold, and dryness in the external world are, in the invisible world of man, fevers, epilepsies, hydropsy, catarrh, paralyses, and apoplexies." Most exciting, or disturbing, depending on who was listening to Hartmann's oration, was what Hartmann had to say next. This new sort of university physician would also have needed to be skilled in the theory and practice of alchemy. "For this one lamp of Diana has," he announced to the university audience, "revealed more than by all the regular physicians combined." Galen and his followers knew nothing of the alchemical art. Hippocrates, on the other hand, understood that medicaments were made by separation and this was what the "skilled natural physician" had also to learn by a marriage of Vulcan and Pallas (in other words, the fire/furnace and wisdom). Knowing these things, the "skilled natural physician" could influence nature by means of nature, examine the secrets of things by means of the fire, and consider the world in the world of man (Moran, 1991).

So, where was all this to be done? What kind of classroom did Hartmann have in mind? The Marburg student could become a skilled natural physician only in one place, in Hartmann's "public chemico-medical laboratory." Hartmann's laboratory was designed to be a fixed space, in contrast to the transitory appearance of many alchemical workshops; and gaining access to that space required students to agree to specific responsibilities and to a prescribed relationship with their teacher. The first requirement was that all students take an oath swearing to their teacher obedience, loyalty, diligence, discretion, and gratitude. Students were to see to the protection of their clothes by providing themselves with a little skirt or apron. They were encouraged to look at everything and to ask about the processes underway. No one was allowed to take anything from the laboratory without the knowledge of the instructor. Those in attendance were to observe the types of chemical utensils and the construction of the ovens. They were also to write down the ingredients used in preparations and to note especially the amount of heat required in different procedures and the lengths of time materials remained in the fire. For his part, Hartmann promised, along with many other recipes, to demonstrate preparations made from opium and to reveal how to make the famous "drinkable gold" of the English alchemist Francis Anthony. He also promised to explain individual chemical terms and phrases, assuring his students that he would repeat anything that was not clear the first time around in order to achieve a "clear, full, and complete setting forth of the facts."

Although he was working in a public laboratory, it is clear that Hartmann considered the knowledge imparted there to be privileged. What students saw, heard, or experienced as a result of their work, he instructed, must never be divulged through any sort of public writing. Hartmann was still concerned to protect secrets, but these were not secrets in the sacred sense. These were trade secrets—secrets that were knowable not through divine inspiration but by means of correct procedural instruction. From the view of someone like Libavius, however, any emphasis on secrecy was unacceptable to chemistry's presence within the university. At Marburg, *chymia,* taught by a Paracelsian, had become something institutionally privileged, private, and unique. For Libavius, this was a horror. If chemistry were to become suitable to the university, it would do so not by becoming anything new or unique but by adapting itself to the procedures of medieval alchemy and traditional, Aristotelian natural philosophy. Academic chemistry, according to the plan he devised, was really public alchemy.

✳ As we have seen, part of Hartmann's laboratory instruction focused on recipes found in Oswald Croll's *Royal Chemistry.* It was, in fact, partly through Hartmann's notes to a later edition of the text that Croll's work became familiar to chemists thereafter. At the beginning of the eighteenth century, a chemist, physician, and professor of the University of Leyden, Hermann Boerhaave (1668–1738) (easily the busiest professor in Europe, holding three of Leyden's five chairs in natural science), began his lectures in chemistry with a list of the books that he believed had added most to the discipline. The list tells us much about a didactic tradition influenced by Hartmann that continued, despite Libavius's reservations, to affect chemical teaching for years to come. Among texts that had "digested the operations of chemistry into a regular system," the first that Boerhaave mentioned was Croll's *Royal Chemistry* with notes by Hartmann. Second on the list was a well-known chemical manual called *First Voyage in Chemistry* or, maybe better, *Apprenticeship in Chemistry* (*Tyrocinium chymicum*), written by a member of the French court, Jean Beguin. Hartmann, whose *Medical-Chemical Works* Boerhaave put third on his list, also edited and added notes to Beguin's text in 1618, preferring for some reason to use the name Christopher Glückradt on the title page (Boerhaave, 1735: vol. 1, 17).

Like Hartmann's courses at Marburg, Beguin's *Apprenticeship* was also a private matter, at least in its original form. Beguin had put together a pamphlet for the personal use of his students and, in 1610, had received all the copies of the pamphlet from the printer—or so he thought. Indeed, printers or printers' apprentices knew a good thing when they saw it. Somehow a copy of Beguin's text came into the hands of a publisher in Cologne where, greed being part of the human condition, it was pirated and printed anonymously in 1611. In response, a very unhappy Beguin resolved "no longer on account of human envy to bury and hide the talent entrusted to me by the wisest and greatest God, but to put it to interest and usury by teaching and instructing students and eager learners in chemistry, and to encompass the whole subject in my writing" (Patterson, 1937: 252–253). Beguin had decided to go public and expected to be paid for the effort.

The decision to go public led to a revised text, including extra chapters, that was published at Paris in the following year (1612). Critics, Beguin knew, would think that he had broken the chemical faith and had divulged secret things. Well, "let them split their sides," he wrote, "let them complain that the greatest injury is being done to the secret philosophy of separation, and let them assign me therefore to all terrible fates for that reason, only let me obtain my object . . . and show the way to truth to those in error, and confirm in the truth those not in error" (p. 253). Some, who had not been able to obtain "by theft and other tricks" what he had now offered to the world would, he believed, repudiate his medicaments. Others would be ungrateful and never acknowledge that his text was the source of their own ability. Human beings were so unscrupulous. But before we begin to feel too sorry for poor Beguin, it is good to remember that thorough citation was not a professional requirement in seventeenth-century writing. Beguin's text, in fact, has a great deal in common with another book, the earlier *Alchemy* of Andreas Libavius. In fact, passages describing laboratory techniques and certain processes were taken over word for word into Beguin's manual—perhaps directly borrowed or maybe to be explained by a text common to both authors (Patterson, 1937; Kent and Hannaway, 1960).

In defining alchemy—or, as he also calls it, chemistry—Beguin decided that it was "the art of dissolving natural mixed bodies, and of coagulating the same when dissolved, and of reducing them into salubrious, safe, and grateful medicaments" (Beguin, 1669; rept. 1983: 1). So alchemy, or chemistry, was something to do, or to make. The subject was the discipline of processes, and this was very different from the way chemistry came to be defined later on. When Boerhaave began his lectures at Leyden, he defined his subject as an art "that teaches us how . . . bodies . . . may, by suitable instruments, be so changed that particular determined effects may thence proceed, and the causes of these effects understood" (Boerhaave, 1735:

vol. 1, p. 19). Boerhaave's concern was for understanding the causes of effects. Learning the practical, procedural stuff was necessary, but not the aim. Nevertheless, Beguin's definition of chemistry as the art of making something is very important, because it is in the act of making that one gained a particular insight into the thing made, a kind of "maker's knowledge." But more on that later. Just as significant for our purposes is something else that Beguin clearly acknowledges—no matter what the teaching of such processes needed to be called in order to satisfy school officials, it was on the learning of artisan alchemists that instruction in chemistry was based.

People, Beguin writes, are really deceived when, "hearing the name of alchemist, [they] conclude, that [this] man employs himself in nothing else than the metamorphosing of metals, and meditates on no other thing than the Philosophick Stone. Whereas the intention of this artist is to prepare most sweet, most wholesome, and most safe medicaments" (Beguin, 1669; rept. 1983: 2). If a kind of metamorphosis took place in alchemy, it was a transformation from that which was poisonous to the body to that which was an antidote to illness and disease. "For if the venomosity of metals and minerals depend upon their form; who sees not," Beguin continues, "if these by chemical artifice be resolved into their three principles, that their deadly and destructive qualities are removed?" By dividing a substance into its constituent parts, the alchemist could both identify and remove that part which was poisonous to the body (pp. 5–6). Even at the beginning of the eighteenth century, one did not have to look far to find the same view of alchemy expressed in the works of authoritative writers. Boerhaave, although in other respects a hardheaded chemical mechanist, found room to note in his chemical lectures that, whether the study were called chemistry or alchemy, those who first used the terms meant nothing more than the pursuit of a universal knowledge of nature. "The word therefore was used originally in a very pure sense, though it was afterwards perverted to a very different one, which misfortune," Boerhaave surprisingly remarks, "likewise befel the word magic" (Boerhaave, 1735: vol. 1, 5).

✳ In his review of chemical writers, Boerhaave created a category for authors who had excelled at applying chemistry to medicine and natural philosophy. The first in this list was Jean Baptiste van Helmont, who incidentally Boerhaave also recommended among those "of greatest repute" in alchemy—so much once again for clear borders between disciplines. Following van Helmont, he recommended Robert Boyle, "in all his writings," and the *Chemical-Physical Dissertations* (1696) of the Leipzig professor Johannes Bohn (1640–1718). He also gave specific mention to the *Foundations of Dogmatic and Experimental Chemistry* (1723) of the professor of medicine at Halle and court physician, Georg Ernst Stahl (1659–1734). Pride of place however was reserved for the text of someone else, the *Physico-Chemical Observations* (1722) of the German professor of medicine also at Halle, Friedrich Hoffmann (1660–1742), "a gentleman who has done a vast deal of service to the chemical art, and enriched both chemistry and physic [medicine] with an abundance of beautiful observations" (pp. 17–18).

The remainder of this chapter pays attention to some of these texts (as well as to others that Boerhaave did not mention) as a way to engage three issues that further connect alchemy/chemistry to the process of the Scientific Revolution. First, we are going to see chemistry become more firmly established as a discipline capable of being taught at schools, particularly in the writings of two authors, Christofle Glaser and Nicholas Lemery. Second, we are also going to consider how theoretical debates—in particular, a debate concerning the role of acids and alkalis in the functioning of the body—focused attention once more on the question of how best to define the principles of nature. Finally, in the same discussion, we are going to be witnesses to another sort of process: The process

by which the earlier, all-embracing subject of alchemy (in which
chymia was a component mostly related to the making of medicines) came to be transposed into an insignificant part of a very
large subject, namely, chemistry. Keep in mind that these three de-
velopments were going on all at once. Our job, in the short sketch
to follow, is to get them untangled.

One of Boerhaave's predecessors at the University of Leyden, a
prominent participant in chemical discussions in the later seventeenth century, was Franciscus de le Boë Sylvius (1614–1672).
Sylvius had practiced medicine both at Hanau and Amsterdam
before joining the medical faculty at Leyden in 1658. From that
position he represented both experimental anatomy and medical
chemistry, and he made use of chemical explanations to describe
both the nature of disease and the functions of the body. Like van
Helmont before him, Sylvius gave his attention to fermentation and
concluded that the process of fermentation was essential to the
physiological process of digestion. Van Helmont had also suggested
that the fermentation that accounted for digestion, although due
ultimately to a spiritual force, was also affected by the operation of
acid in the body and, indirectly, by the presence of alkali. In that
muted moment, a new view of the basic components of the chemical operation of the body was born—the acid/alkali theory. What
bubbled up from this theoretical mixture continued to interest
chemical writers for several decades thereafter.

Prompted by ready-to-hand observations of violent reactions
occurring when mineral acids combined with alkaline substances,
Sylvius began to think that not only chemical processes could be
explained by acid/alkali reactions, but that also diseases themselves
resulted from the acidic and alkaline natures of specific bodily
fluids (for example, lymph, saliva, pancreatic juice, and bile). The
turbulence and strife between acids and alkalis in the body was de-
tectable, he argued, by the presence of effervescence. Because diseases were caused by an overabundance of acidity and alkalinity,

these could be neutralized by the plop-fizz of acidic or alkaline medicines of the opposite sort. An acid stomach found itself neutralized by an alkaline solution—Alka-Seltzer™ without the brand name.

By the later seventeenth century, van Helmont's acids and alkalis had come to be regarded not just as certain sorts of chemical substance but also as opposing chemical principles, the agitation and strife between which accounted for all chemical and physiological reactions. Sylvius's student Otto Tachenius was so enamored of the idea that he claimed that acid and alkali were the "architectonick instruments of nature" found in all sublunary things. From this perspective, the theory of acids and alkalis turned into an alternative theory for the four elements of Aristotle and the three chemical principles associated with Paracelsus. It also proved easily adaptable to the assumptions about matter being made by representatives of the mechanical philosophy. Before we talk more about that, however, we need to check in once more with other attempts to bring chemistry into public view (Boas, 1956).

✳ The tradition of teaching chemistry established in Paris by Beguin continued there in a setting that seems strange at first glance, a location known as the Garden of Plants. Although the garden, which was designed as a royal garden of medicinal plants, was not officially founded until 1635 (and not officially opened until 1640), the first draft of the project included an important teaching component in which a resident druggist was expected to offer instruction in the preparation of herbal medicines and distillation techniques. The garden was equipped with lecture halls and laboratories and was augmented further in 1648 by the creation of an official teaching position in chemistry and botany. The man chosen to fill that position, a transplanted Scotsman named Guillaume Davisson (William Davidson) (ca. 1593–ca. 1669), was himself a court physician who had been teaching informal courses at Paris in chemical medicine for over a decade. Laboratory instruction in chemical medicine continued at Paris with Nicolas Le Fèvre (ca. 1615–1669), who succeeded Davisson in 1651, and Christofle Glaser (died ca. 1670–1678), an apothecary to both Louis XIV and to his brother, the Duke of Orleans. In charting the didactic experience of chemistry, Glaser is especially important. His text, written in French, the *Treatise Concerning Chemistry* (1663), was a direct product of organized teaching and focused far more attention on practical procedures than had the written works of either of his predecessors.

Glaser knew that alchemy and chemistry had become interchangeable terms for many, but recognized "chymistry" as the term most in use when defining "a scientific art teaching how to dissolve bodies, how to draw out from them the different substances of their composition, how to unite them again, and how to bring them to a higher perfection (Glaser, 1663: 4–6). Much like Beguin, therefore, Glaser saw chemistry as the description of process. It offered the means of entry into nature's secrets because it described how to reduce things into their first principles and how to give them new forms. Chemistry accounted as well for the new rationality of the human body in which fermentations, digestions, circulations, corruptions, separations, distillations, and other chemical operations explained essential vital processes and helped to elucidate the functioning of each of the body's parts. Chemical processes maintained health, but they were also responsible for occasions of illness. In this last respect, chemistry provided knowledge of the nature of disease and at the same time supplied an understanding of the most effective remedies to combat disease. Apothecaries, Glaser told his students, relied on chemistry to teach them how to make compositions, how to preserve the virtues of their ingredients, and how to separate the pure from the impure parts of mixtures.

But chemistry furnished the basis for other forms of knowledge as well. Its practical benefits for painters, engravers, dyers, and other craftsmen were beyond dispute. (pp. 1–4). After all, the subject of chemistry was vast, embracing the animal, vegetable, and mineral worlds. Boundless too were the possibilities for theoretical speculation, and Glaser dispensed to students not only information about chemical processes but also his own thoughts about the constituents of matter. He noted that all things could be reduced by the fire to five first principles—three active principles (a Mercury that was spiritual, a Sulphur that was oily, and Salt that was, well, salty) and two passive principles (watery phlegm and earth). While the principles remained mixed in a body, the virtues of the active principles remained hidden. Chemistry, however, separated them, purified each, and united them again so as to create purer and more active substances (pp. 7–8).

The textbook of another royal apothecary, *A Course of Chemistry* (1675) written by Nicholas Lemery (1645–1715), went even further. In fact, if we are looking for a place where "alchemy" was redefined and discarded in favor of "chemistry," we can find a good candidate here. Lemery wanted the subject he taught to receive the blessing of the French academic community, which was mostly composed of critical sorts who, like the philosopher they revered most, René Descartes, wished learning to proceed from a skepticism of all things received from the past that claimed to be true. To make a clean break with previous interpretations of nature, Lemery cast alchemists into the ranks of frauds and imposters who were (all of them) solely concerned with making gold. Redefining alchemy in this way allowed chemistry to shed any connection to dubious alchemical practices. Chemistry was laundered so as to have an untraceable history. By virtue of its shared methods and types of inquiry, it claimed to be a distinct and unprecedented form of knowledge possessing its own rational mode of discovery. The new perception of chemical experience excised perceived alchemical lies and deceits and turned what had been practical alchemical wisdom into new chemical facts. Alchemy had entered a phase of cultural metamorphosis. In that state the *Course of Chemistry* became a publisher's dream. It went through thirty editions by the mid-eighteenth century, and you could read it in Latin, German, Dutch, Italian, Spanish, or English (Powers, 1998).

Of course what one read about in the *Course* would hardly be recognizable by anyone today. Lemery presented a system of chemistry based on the mechanical philosophy, and one that recast the acid/alkali theory already discussed by van Helmont, Sylvius, and Tachenius to fit a mechanical model. Acid particles Lemery described as possessing sharp points, which penetrated the porous bodies of alkali particles when the two were brought into contact. The resistance caused a ferocious bubbling, or effervescence. During a chemical reaction, the acid points broke off and were "blunted" inside the alkali pores, forming a salt. There were lots of salts of different kinds, depending on the makeup of their acid and alkali constituents; and ultimately Lemery advanced the view that all substances, including metals, were made up of various compounds of acids and alkalis. The model accounted for a wide range of chemical phenomena. For example, Lemery explained the fact that some acids would not react with certain alkaline substances by positing that the points of these acids were of improper size or shape to penetrate the pores of the alkali. Similarly, the bubbling that occurred when a fixed alkali was added to an acid solution was caused, he claimed, by the dislodging of particles of fire that remained in the pores of the alkali after its synthesis through the combustion of plant matter (Boas, 1956; Powers, 1998).

It was a nice theory and, even though uncontaminated by alchemical speculation, one that would hardly last the century. Moreover, taking Lemery's mechanical interpretation of the acid/alkali theory at face value and characterizing his description of matter as a radical break from previous chemical approaches is, some have claimed, not only to undervalue the persistence of preceding alchemical ideas but to distort the extent to which the mechanical philosophy itself was a driving concern in his chemistry. As the historian of science Jan Golinski reminds us, chemistry does not have to reduce natural phenomena to matter and motion to be relevant to the Scientific Revolution (Golinski, 1990). In fact, for Lemery mechanical descriptions may not have been indications of unconditional philosophical allegiance as much as heuristic attempts to describe chemical operations in a simple (mechanical) way so that students could more easily form a picture of what was happening during chemical processes. Ultimately, Lemery had to concede that the principles of things, although sensible, only really existed as a necessary invention to aid explanation. "The word principle in chymistry," he wrote, "must not be understood in too nice a sense: for the substances which are so called are only principles in respect to us; and as we can advance no further in the division of bodies . . . [but] . . . we well know that they may be still divided into [an] abundance of other parts which may more justly claim . . . the name of principle" (Lemery, 1698: 5–6).

The structure and content of Lemery's *Course* was very similar to Glaser's *Treatise*. In fact, early editions of Lemery's text described chemical operations and pharmaceutical recipes identical to the ones found in Glaser's text. Like Beguin and Glaser before him, Lemery refused to think of chemistry in the Paracelsian sense as the purification and manipulation of essences, but rather as "the art of separating different substances that are encountered in a mixt." "Mixt" meant the compound state of naturally occurring bodies; and Lemery accepted, as Glaser had also, that "mixts" were composed of five chemical principles: spirit, oil, salt, water, and earth. Others had made similar attempts to redefine the principles of bodies, and one that is particularly noteworthy at this point was advanced by an English physician and professor of natural history at Oxford named Thomas Willis (1622–1675) (Debus, 2001: 57ff).

To Willis, Aristotle's elements (earth, air, fire, and water) needed to be rejected because they provided no special insight into "the more secret recesses of nature." The atomist philosophy of the ancient Greek thinkers Democritus and Epicurus, on the other hand, deserved praise, he thought, for endeavoring to explain natural phenomena "without running to occult qualities . . . and other refuges of ignorance." Nevertheless, Willis had to admit that the atomists often just presumed rather than demonstrated their principles, and that their notions were very remote from sense experience. His own view, that all bodies could be resolved into particles of Spirit, Sulphur, Salt, Water, and Earth, at least had the merit that the principles of things were sensible and could be discovered by means of the process of separation (Willis, 1681: 2).

For both Willis and Lemery, the essential thing was that the "principles" of matter, if not perhaps the most fundamental reality, were nevertheless basic material substances and not Aristotelian qualities or spiritual presences somehow rooted in matter. Especially Lemery, in this regard, could have his cake and eat it, too. By insisting that chemical principles were "sensible" and "demonstrative," he preserved a traditional way of describing chemical operations in iatrochemical terms (acids and alkalis) while, at the same time, he allowed for the mechanistic and materialistic explanations that Cartesian philosophy demanded (Powers, 1998). That kind of a "mixt" truth extended also to a mixing of interpretations, both academic and popular, when it came to appraising the further status and reputation of alchemy (Figure 7).

As we have seen, Lemery felt it was his moral mission to protect the public from alchemical tricks and, as a result, redefined alchemy as simple gold-making. Nevertheless, the very public that he sought to shelter was not so willing to give up altogether on alchemical experience. In fact, a great portion of the reading public was still eager to gulp down, ironically under Lemery's own name, large doses of what had long been considered alchemical secrets. It is not altogether certain who wrote the *Collection of Rare and New Curiosities* that appeared first in French in 1674. The book was a hot seller,

Figure 7. Mid-eighteenth century depiction of various distillation instruments and processes within a pharmaceutical laboratory. The shop is visible through the doorway. The books on the shelf are inscribed, from top to bottom, "Paracelsus, Pharm[acopeia] Lo[ndinensis], Boerhaave, Lemery." Wellcome Library, London.

however (five French editions over the next fifteen years and two further printings in the early eighteenth century). The Library of Congress attributes the book to Lemery, while the British Library claims it is the work of someone with a very similar sounding name, Antoine d'Emery. Regardless of who wrote the French text, by the time the book was translated into English, Lemery had written it. And despite his claim to fame in academic circles as a rational mechanist who thought all alchemy was nonsense, Lemery met the public, in this text, as a dealer in secrets. What the public purchased was, in other words, not strictly a book about *la chymie*, but a text that most would have recognized as part of a long alchemical tradition—a text that advised how to destroy bugs, make ink, polish brass and silver, whiten teeth, keep roses fresh, take spots out of silk, and (my favorite) get rabbits out of the berries without using a ferret (Lemery, 1685). From the public point of view, alchemical experience still mattered, even if dressed up as the latest chemical thinking.

✳ Not everyone thought of alchemy in the same way as Lemery did, and not everyone was as eager as Lemery to embrace the acid/alkali theory of matter. The famous experimentalist Robert Boyle disliked the language of "strife" used by supporters of the hypothesis and thought that the idea of relying on effervescence to determine the presence of acids and alkalis was vague and uncertain. One of Boyle's biggest fans, a German professor named Johannes Bohn (1640–1719), also declared acid and alkali to be insufficient to serve as the principles of natural bodies. In chemistry he preferred an even more "skeptical path" in explaining the number and nature of natural principles. "I do not deny," he wrote, "that acids and alkalis perform powerful reactions in chemistry," but this did not mean that they should achieve the status of chemical principles. Moreover, just because the elements of Aristotle and the three principles of the Paracelsian chemists had been challenged and rejected, there was no need to flee to other assumed precepts and invent acids and alkalis in their stead. In fact, "saying that everything effervescing with acid is alkali and everything bubbling with alkali is acid" was to draw a conclusion on the basis of observations that were altogether ambiguous (Bohn, 1696: 523).

Neither Boyle nor Bohn put an end to thinking of acids and alkalis as fundamental principles or elements, however, and theories linked to acids and alkalis continued to surface for several years to come. The German chemist Georg Stahl (1659–1734) argued for the existence of a "universal acid" and considered salts to be mixtures of this acid with one or more of three kinds of earth. Stahl's ideas influenced the thoughts of another experimentalist, Wilhelm Homberg (1652–1715), who brought to bear quantitative techniques on the acid/alkali problem. In two papers published under the auspices of the French Royal Academy in 1699 and 1700, he measured the relative strengths of acids and alkalis by chemical and physical methods. Remarkably, given the growing hostility toward alchemical claims, he also managed to merge a view of matter made up of tiny parts, or corpuscles, with the medieval Sulphur-Mercury theory of metals while claiming that he had successfully experimented with making gold using Philosophical Mercury—a good example that not everybody doing chemical science followed Lemery's urging to separate chemistry from metallic transmutation (Principe, 2001).

Referring to acids and alkalis in order to understand chemical reactions was one thing, but thinking of acids and alkalis as basic to matter, or as principles of matter, was clearly another. Hermann Boerhaave accepted the former application and knew of Robert Boyle's technique of identifying acids, alkalis, and neutral substances by observing changes in color when a substance was dipped onto a little "syrup of violets" spread on white paper. However, in a work called *A Short Recapitulation of Acid and Alkali* that became part of his *Elements of Chemistry*, he noted how thinking of acids and alkalis as fundamental parts of matter had, by the 1730s, already become a thing of the past. "Some of the greatest men in the art [of chemistry] have been guilty of this childish error," he noted; but "how trifling is the calling of the assistance of alkalis and acids to explain all the phenomena of natural bodies? And yet we have seen the time when this doctrine was so much in vogue, that it was thought an honour of the age which entertained it" (Boerhaave, 1735: vol. 2, 374). Like the alchemical theories that Lemery despised, the acid/alkali theory that he endorsed had had its day, and it even seemed ridiculous in retrospect.

✳ Johannes Bohn had been moved to write his *Chemical-Physical Dissertations* (1696) by a bookseller in Jena who encouraged him to bring together his disputations and chemical notes as a way to confront the errors and prejudices of the chemical art on the basis of more accurate observations and experiments. In the preface, he noted that if Hermes, Geber, and Lull had come back to life, they would not have recognized the many new distillations, circulations, and calcinations that had been refined over the years but that were nevertheless derived from their own writings. The interesting thing is that Geber, Lull, and others were no longer referred to as alchemists. They had become "chemists"; and chemistry, Bohn decided, was made up of four parts. The first was a philosophical part that concerned theorizing about the principles of natural bodies. The second part was pharmaceutical and involved the preparation of helpful remedies. The third part he called "mechanical" and defined it as having to do with things that were artificially made or contrived by beer makers, dyers, glass makers, soap makers, metallurgists, goldsmiths, and similar craftsmen. The fourth part was alchemical with the solitary aim, he observed, much like Lemery, of the transmutation and exaltation of metals. Each of these parts, however, shared something important in common, which was the thing that really defined "chemistry." Each was involved in the process of separating natural mixed things into parts and then, in turn, of recombining these parts with other things to make compounds. In this way, says Bohn, chemistry "inquires into the order and causes of the emerging compounded phenomena: in a word its end is the work itself" (*uno verbo, finis ejus est ipsum opus*) (Bohn, 1696: *praefatio*). In Bohn's view, the end and the means of chemistry were the same. Libavius, Beguin, and numerous earlier alchemists would have said the same. The purpose of their art was in the doing of their art. Whether the doer was the medieval Lull or the more modern Lemery, both, Bohn knew, were doing science.

According to Bohn, the study of chemistry was fundamental to perfecting all the other arts and sciences. No one who wished to be successful in medicine could ignore chemistry, and the real professors of natural philosophy in our age, he noted, were "men of the body" who understood the texture, nature, and structure of the body's various parts. Most of all, however, the universality of chemistry consisted in the fact that it alone displayed the means by which mixed bodies were dissolved and their textures transformed. Chemistry could thus alter the innate properties of bodies and direct them into other things. It is incredible, says Bohn, "how much power the chemist has." It was "this noble and excellent part of philosophy" that Bohn had "loved . . . since boyhood." What he really loved, however, was the power to make things different than they were before, to force nature, as it were, into different shapes and structures, and from that to learn what was fundamental to her construction. In this, as we shall see, Bohn shared an important attitude toward the creation of knowledge that had recently been expressed in a more philosophical setting as the experimental method.

While Johannes Bohn was busy compiling his notes at the University of Jena, a place for chemistry was also being prepared at the Dutch University of Utrecht, and here too chemistry was featured as a type of skill with which to impel changes in nature. The person who shaped the physical and intellectual space for chemistry at

Utrecht was Conrad Barchusen (1666–1723). Treating what he called "Pyrosophia" (knowledge of the fire), Barchusen noted that a chemist had the power to create what appeared to be entirely new material beings. Actually, what the chemist did was to dismantle, by means of different regimens of the fire, "as if by a first rate instrument," the parts of substances that were formerly joined together in one kind of body and then to recombine them so as to make something different. The change was often so thorough, however, that one could think that the chemist had brought into being something that in no way had existed before. Analysis and synthesis were therefore the preoccupations of the chemist and, in teaching how to do both, Barchusen divided the subject of chemistry into three parts, each with a didactic purpose. The first part was *iatro-chemia* or medical chemistry. This was the art of teaching how to prepare medicaments from different bodies. The second part included instruction in the various ways to compound metals and knowledge of the numerous secrets of metallurgy. The third part of chemistry was what Barchusen called *alchemistica,* or the Hermetic art that concerned, he wrote, transmuting cheaper metals, like iron and lead, into more precious metals, like gold (Barchusen, 1698: 4–5).

Clearly Barchusen did not include this last aspect of chemistry in his university instruction, and he acknowledged that his understanding of it was by hearsay only. Nevertheless, what he had heard about *alchimistica* had been communicated to him by people eager for the truth. Experts (as well as frauds) agreed that Paracelsus *and* Libavius had achieved the transmutation of iron into copper, for instance. Moreover, even if alchemy was defined solely as having to do with metallic transmutations, there were two ways, he reported, to pursue this end. One way was by means of possessing universal knowledge, which required the adept to seek the Philosophers' Stone through enigmas and parables. Clearly this strategy had nothing to do with chemistry, and Barchusen lets us know exactly what he thinks of it. "Good god" *(proh Jupiter),* he exclaims, this was knowledge based in fables and whatever spectacles might be claimed as a result of such insight could never be reproduced by laboratory methods. But, besides the universal approach, Barchusen conceded that there were other, more particular ways in which transmutations might occur; and these, he insisted, were based in method and specific laboratory techniques (pp. 422ff).

Although metallic transmutation did not take place within the laboratory at Utrecht, we still have a good idea of what sorts of procedures occurred there. We know of them because one part of Barchusen's book *Pyrosophia, Examining Concisely and Briefly Iatrochemical and Metallic Matters as well as the Business of Making Precious Metals* (1698) stands out from the rest of the text as a kind of advertisement for his course in chemistry. This section of the book, added as an appendix, is called "A review of the chemical labors in the second semester of the year 1695 in the academic laboratory at Utrecht." And it is there that Barchusen declares that his general purpose in constructing the course was to demonstrate for students how "sublunary bodies can be reduced into four different substances or principles: namely salt, oil, water, and earth; and [how] these [can be] examined within various mixtures and combinations by the work of different fires and procedures" (p. 445). In other words, the procedures that students learned helped them discover the sensible and manipulatable constituents of matter. Among other processes, students in this particular semester learned to make use of the important methods of distillation (as a way of resolving bodies into their principles), of incineration (so as to furnish the fixed salt of alkali), of putrefaction (in order to produce the volatile spirit of urine), of coction and inspissation—in other words, thickening by evaporation (to exhibit salt of tartar), of fermentation (in order to display how it creates a "burning spirit"), and of the means to bring forth fragrant essential oils (how to make perfumes). The same procedures could be followed to discover the chemical compositions of animate substances, and students in the

Figure 8. Conrad Barchusen's laboratory at Utrecht. Barchusen's laboratory was a "passive" instrument containing other "active" instruments for the purpose of making new things. *Pyrosophia* (Lugduni Batavorum, 1689). By permission of the Syndics of Cambridge University Library.

Utrecht laboratory experienced how to apply them to the examination of the components of blood, urine, and dung. Minerals had their parts and principles more intricately bound within them; but Barchusen promised to examine these as well, showing to students how to resolve them so as to produce their salts (pp. 445ff) (Figure 8).

Theory followed from procedure in Barchusen's laboratory. Knowing how to use instruments in unlocking the parts of mixed bodies was especially important; and, in an earlier part of the *Pyrosophia,* Barchusen observed that some of those instruments should be regarded as "active" and others as "passive." Those labeled passive were instruments that did not predetermine a particular kind of operation but simply allowed things to happen (*sed modo patiuntur).* The most important of these was the laboratory itself, and Barchusen gave attention to how the laboratory space should be organized. He required that it not be too narrow and that it be placed in a salubrious location through which air could enter and exit freely in all directions. There needed to be amble room for various procedures involving fire, and there was also to be a cistern with fresh water ready at hand. Within this "passive" instrument, or laboratory, there were to be found a number of "active" implements, and the purpose for these was to force things to occur. Among the most important were furnaces of various configurations, and Barchusen described these for his readers in detail. The care taken to describe procedures and to clarify the use of instruments was also extended by Barchusen to a precise naming of the active instruments themselves. Because readers and potential students came from different places, and Holland itself was a crossroads of cultures in the late seventeenth century, Barchusen judged it a good idea to designate some vessels and utensils by their Latin, German, French, and "Belgic" names, and to add pictures of instruments (including his laboratory)—just to be clear (pp. 62ff).

✳ Barchusen's "passive" laboratory was designed as a space in which to show how chemistry could change the situation of bodies, rearrange their parts, and, by so doing, provide a special kind of knowledge about how nature herself was put together. To do this, however, required not just passive instruments but active ones as well—instruments that would force nature into relationships in which she was not naturally found. For Francis Bacon (1561–1626), this kind of approach was tantamount to putting nature on the rack and, as we will see next, the production of this type of "experimental" knowledge through the forced manipulation of nature's parts placed chemistry within the company of other disciplines at the end of the seventeenth century that were getting downright pushy in their attempts to gain new learning.

CHAPTER FIVE

ALCHEMY, CHEMISTRY, AND
THE TECHNOLOGY OF KNOWING

There is a difference between experience and experiment, and Barchusen's distinction between passive and active instruments draws attention to it. If you watched a mouse all day long, you would probably never have the experience of seeing it wander into a void space, or vacuum. Something happened in the seventeenth century, however, with its attention given to active instruments and to the process of learning via experiment, that added a new worry to the normal anxieties of mouse life in Western Europe. It became possible to build an apparatus—an air pump, for instance—that artificially created what nature herself had not provided, at least not in the typical living space. The apparatus, or experimental instrument, forced nature into situations not readily encountered through passive experience. Once manufactured, mice, candles, clocks, and a variety of other things might just wind up inside a vacuum jar, put there deliberately in order to test a variety of hypotheses about the air.

Among the first to build air pumps were the English experimenters Robert Boyle and Robert Hooke, but the notion of "experiment" as "the constraint of nature" was already prepared long before them. The idea received a great deal of attention in Francis Bacon's *Novum Organum* (1620), a title that loosely translates as *New Method*. There Bacon argued that in order to learn any of the secrets of nature, one had to be aggressive, one had to put nature on the rack, so to speak, and wring the truth out of her. Real learning occurred not when nature was "free and at large," but when nature was "under constraint and vexed; that is to say, when by art and the hand of man she is forced out of her natural state, and squeezed and moulded" (Bacon, 1620; rept. 1960: 25). Furthermore one needed to follow a particular method of inquiry in which theories arose as a result of collecting and organizing individual observations and natural facts. Most of all, Bacon decried the limits and vanities of established knowledge and called for a new science based on a "commerce between the mind and things" and a "lawful marriage between the empirical and rational faculty." This was exciting philosophy, but haven't we heard before of a need to jostle and shove nature by art so as to create for ourselves what nature had not provided? Oh yes, while not stated as a new approach to learning, something nevertheless sharing in this view used to be called transmutation.

Bacon was skeptical of received opinions grounded in ancient authorities, and in this he shared much in common with the French philosopher, René Descartes (1596–1650). Descartes's *Discourse on Method* remains one of the most important texts in the western intellectual tradition, but not everyone has evaluated it in the same way. In fact, some have argued that its agenda of achieving mathematical exactitude and intellectual certainty through reason and method probably did more harm than good to a Renaissance tradition in which uncertainties, ambiguities, and differences of opinion were at least acknowledged as philosophically inbounds. Moreover, Descartes's demand that philosophy should seek out abstract, general ideas in order to make sense of accumulated personal experience emphasized a mathematical approach to understanding the world—all one had to do was to throw a kind of geometrical grid over nature and compute relationships.

If you want to say, "it's a fact" in Italian, just say *La matematica non è una opinione* (mathematics is not an opinion). Whether or not Descartes knew Italian, he would have agreed with the statement. Theology was, of course, another sort of certainty, but one, in its institutionalized Christian forms, that was based in the revealed word of God. The consequence of providing two ways to be certain was the construction of the well-known Cartesian dualism—two realms of being existing at the same time, but not, except in human beings, ever overlapping. Descartes called one category the *res cogitans* (things of the mind or thought). The other, which corresponded to the physical world, he called *res extensa,* objects that take up space. The certainties of natural philosophy were limited to the latter; and in the investigation of nature, references to two things, and these two things alone, were acceptable as types of physical explanation. The two items were matter and motion. We have already called this the mechanical philosophy. Descartes, without a doubt, was its primary architect.

These aspects of Descartes's philosophy are well known. However, two references in the *Discourse,* which are easy to miss, are equally important to our discussion of how chemistry relates to the process of scientific revolution. Both come in the very last part of the text. The first is a metaphor and the second is a personal reflection. Thinking critically of the kind of knowledge created by means of medieval debate, Descartes compared knowledge based on Aristotle to ivy that could climb no higher than the tree supporting it and that even tended to grow downward again after it had reached the top. The tree, in this case, was the dead wood of ancient authority and Francis Bacon, a few years earlier, had been just as condemning of it. To make knowledge for oneself, one needed to be guided by a rational method. Just as important, however, one needed a new routine, the habit of discovering truths on one's own, "seeking first easy things" and then, by using one's own hands or those of a well-paid craftsman, "passing by degrees to more difficult ones." If knowledge of nature were to be gained, one had to be prepared to make it on one's own. Says Descartes: "I am convinced that, had someone taught me from my youth all the truths for which I have sought demonstrations, and had I had no difficulty in learning them, I might perhaps have never learned any other truths, and at least I never would have acquired the habit and faculty I think I have for finding [them]" (Descartes, 1637: part 6).

✳ The habit of finding truths through demonstration clearly applies to the methods of the seventeenth-century chemist. However, while busy making knowledge by means of experiment, physicians and natural philosophers still found ancient authority important and often invoked it, albeit in new ways, to support their views of nature. When the Helmontian iatrochemist, Otto Tachenius, for instance, wanted to buttress his claims for acid and alkali as principles of matter, he gave credit for the discovery to the ancient physician Hippocrates. Hippocrates, he admitted, used different names, calling the principles fire and water instead of acid and alkali. Nevertheless, Tachenius insisted, it amounted to the same thing because Hippocrates had claimed that "these two [water and fire] . . . can do all things, and that all things are in them" (Tachenius, 1677: "Clavis," 2). So, Hippocrates, the ancient physician, is reborn as a modern chemical thinker, and Tachenius's most popular book arrived at the bookstore with the befitting title *Hippocrates the Chemist.*

Even if you just leafed through the text, you probably would get a chuckle from Tachenius's reference in the preface to two figures, one an old woman and the other an old man. The old woman speaks for popular opinion and holds chemistry in great esteem for providing the means by which she can color her hair. The old man turns out to be Hippocrates himself, and he has something more philosophical to say about the chemical art. Without chemistry, not only would one crave good hair products, but also one would be deprived of all the other arts. Chemistry, he insists, is the source and mainstay of them all. "Whatsoever famous and excellent thing is performed by art," old man Hippocrates says, "it proceeds from the foundation of this ancient [chemical] philosophy." Men know this well enough, he continues, "yet they are ashamed to speak it out." That is certainly saying something. Chemistry, long viewed as a ragamuffin among disciplines, appears now as the parent of them all. Tachenius, however, was not quite finished. An even more stunning observation comes next. "The divine mind," the figure of Hippocrates confides, "has instructed men to imitate her works; they know what they do, but are ignorant of what they imitate" (Tachenius, 1677: "Hippocrates," preface). Learning, in other words, comes through doing. Learning comes about by imitating nature, even though the ultimate reasons nature has come to be as it is reside only in the mind of God.

A few years later, the Italian philosopher Giambattista Vico (1668–1744), driven by much the same thought, would put it this way: What is true is precisely that which is made. In Vico's perception of things, the human being was so integrated into the surrounding world that she could achieve a "witnessing consciousness" of the divine through the process of making things. Real knowledge is what we know of ourselves as human beings through things that we directly make—art, mathematics, societies, and so on. As far as natural knowledge is concerned, however, the things we make, and therefore the truths we declare, always imitate God, who, as the maker of all things, possesses the real knowledge of what He makes (Vico, 1994). Yet, even though our knowledge of nature can only imitate what has already been made, we can nevertheless accept natural knowledge as a process, a kind of technology—not a technology defined as apparatus, however, but a technology that is expressive of know-how. Maker's knowledge in this sense is not a metaphysical category, but a logical one (Goetsch, 1995: 15–17). By intervening in nature through experiment, Francis Bacon had already argued, we entertain a process and produce an effect, one that might be used for the benefit of society. This amounts to genuine knowledge.

If we are looking for an address where we might find under one roof representations of maker's knowledge, experimentalism (expressed in the Baconian sense as the "constraint of nature"), chemistry, alchemy, and the mechanical philosophy, we only have to check in at the house of Katherine Boyle on London's Pall Mall, even then one of the town's more ritzy neighborhoods. There her brother Robert had a laboratory and what was going on in that laboratory and at several other locations where he worked for almost forty years is one of the best examples of how alchemy and chemistry relate to the process of the Scientific Revolution.

Robert Boyle has frequently been described as an important advocate of the mechanical philosophy and as a model of experimentalism (Boas, 1976). Recently, however, he has gained the attention of scholars who have sought to portray him, along with other early modern notables like Galileo, less as heroes of modern science than as true historical figures operating within worlds very much different from our own. Such scholarship has also not been afraid to point out the personal social and political agendas of some of those who had long been considered icons of the Scientific Revolution. In this light, a new image of Boyle has begun to emerge, one depicting him as using experimental philosophy to pursue personal political advantage (Shapin and Schaffer, 1985) while, at the same time, using his social standing as an English gentleman to gain the trust of influential friends in behalf of his scientific claims (Shapin, 1994). In addition, we now have books that describe Boyle's connection to medicine (Kaplan, 1993) as well as his association to alchemy (Principe, 1994, 1998). Most of all, however, we have the image of a complicated and self-entangled Boyle, a Boyle who accepted revelation as well as reason, who was acutely concerned for his reputation (to the point of semi-plagiarism and a denial of influence), and whose religious preoccupations became converted into experimental intensity and punctilious intellectual fervor (Hunter, 2000). In other words, we have got a real human being on our hands, not a scientific fetish.

Part of being human for Boyle meant that the ultimate causes of things, God's knowledge in the act of making them, could never be known. Boyle placed limits on the use of reason in theology and those same limits operated also for him in the realm of natural philosophy. "His God," writes Jan Wojcik, "had deliberately chosen to limit the power and scope of human reason, leaving human beings in something of a state of perpetual blindness concerning the ultimate truths of both nature and Christianity" (Wojcik, 1997: 212). On the other hand, knowledge as know-how—in other words, knowledge gained in imitation of nature—could lead to God. In his *Christian Virtuoso* (1690) Boyle notes that the knowledge of nature might be adapted to oppose or defend religion. In the hands of the atheist or "sensual libertine" (in other words, a convinced materialist), it could be used to discredit religious practice. But it would be different, he adds, if such knowledge were to come into the hands of "a man of probity and ingenuity," or at least one "free from prejudices and vices." Then the improvement of the truths of philosophy would guide one's sentiments of religion (Boyle, 1690: 6–7). Therefore, he concludes, if any of those who cultivate real philosophy (a philosophy that people also called "new, corpuscularian, atomical, Cartesian, mechanical") would use it to countenance atheism, "tis certainly the fault of the persons, not the doctrine" (p. 9).

Cartesian (mechanical) thinking did not necessarily lead to atheism. In fact, Boyle argued that Cartesian principles could actually serve to defend the presence of divine providence in nature. Something, after all, had to account for the continuation of the peculiar motions of the world, and had to guide the action of particles into becoming this or that individual thing (pp. 34–35). Because Cartesians believed that the rational soul, as an immaterial substance, was distinct and separate from the body, what was left in nature as a guiding or organizing principle, but the hand of God? "Whence I infer that the divine providence extends to every particular man . . . since I understand not, by what physical charm or spell an immaterial substance can be allured into this or that particular embryo" (pp. 34–35). The human being may be part of a world of matter, but she was no mere machine—in fact, neither was nature.

Boyle's early ideas about chemistry can be found in a book called *Some Considerations Touching the Usefullnesse of Experimental Natural Philosophy* (1663). There he describes chemistry, which extracts the more active parts of bodies and enriches the virtues of remedies, as the very backbone of medicine. Like van Helmont and others, Boyle recognized that physiology rested on chemistry and that therefore a knowledge of ferments and an understanding of digestion as a chemical process were pivotal in grasping God's design for the operation of the body's parts. But Boyle went further, mixing traditions based in both Paracelsus and van Helmont with experimental studies.

On the one hand, Boyle rejected the Paracelsian tradition that treated the human body as a microcosm analogous to the structure of the greater world, or macrocosm, and he also rejected the idea that Sulphur, Salt, and Mercury were principles of nature to which mixed bodies could be reduced by means of fire. Nevertheless, Paracelsian residues remained in Boyle's writings, and it is clear that his understanding of nature allowed for the presence of spiritual forces and also for cures by means of sympathetic magic. It was the influence of van Helmont, however, that made the greatest impression; and in this regard a good deal of Boyle's attention, often in the company of an alchemical collaborator named George Starkey, was given to duplicating some of van Helmont's most impressive chemical preparations. Of special interest was the universal solvent that van Helmont called *Alkahest* and which Paracelsus was also thought to have prepared under a different name. According to van Helmont, the *Alkahest* turned everything, including metals and minerals, into their elementary constituents; and for van Helmont, the basic stuff of everything was water. Try as they might, Boyle and Starkey never succeeded in duplicating the Helmontian solvent, even though Starkey at one point claimed to have received instructions in a dream directly from God (Clericuzio, 1993: 314ff).

As much as Boyle remained committed to Helmontian iatro-chemistry, therefore, the problem of duplicating the *Alkahest* became an increasing difficulty in accepting van Helmont's larger claims about the material foundations of matter. Could water really be taken seriously as the fundamental principle of all things? And how could metals and minerals especially be generated from such a watery element? By the time he wrote the book for which he is most famous, the *Sceptical Chemist* (1661), Boyle reasoned that the entire issue was just not subject to investigation because the secret of the *Alkahest* had, apparently, only been known to van Helmont, and van Helmont was in his grave. Thus the claim that metals and minerals were reducible to water, says Boyle, "cannot be satisfactorily examined by you or me" (Boyle, 1661; rept. 1911: 73). Moreover, he continued, even "supposing the *Alkahest* could reduce all bodies into water, yet whether that water . . . must be elementary, may not groundlessly be doubted." In other words, just because water could be extracted from different bodies (van Helmont, remember, had shown that most of a tree was water), and, by means of some universal solvent, could even be drawn from minerals and metals, this was, in itself, not sufficient proof that water was an elementary substance.

Van Helmont, of course, had also thought that, along with water, individual bodies were created by means of the transformative properties of what he called "seeds," or *semina*—not really seed-like things in the literal sense, but powers that organized the original matter into a specific object. Boyle ultimately rejected this, too, and he did so on strictly empirical grounds. Plants and animals did appear to come into existence as a result of some form of seed-like growth; but other things, Boyle decided, whether found in nature or created artificially, were really just mixtures and different compositions of bodies and required no generative principle to account for them. And yet, even with Boyle's skepticism, other seventeenth-century chemists were not quite so willing to throw out the baby with the bath water. The most interesting thing of all is that several chemists in the age of mechanical philosophy found ways to combine Paracelsian and Helmontian ideas of guiding forces, spirits, and alkahests with a basic corpuscularian, or atomistic, view of matter. The problem of explaining how matter knows how to organize itself into specific things with specific qualities, and how the parts of living things know how to function, just would not go away.

In 1671, for instance, the English chemical reformer John Webster published a book called *Metallographia*. The text, which made use of Boyle's *Sceptical Chemist* to support its arguments, declared that metals were generated from seeds guided by the action of a "plastic principle." Atoms, Webster asserted, certainly existed but were really only good for increasing the bulk or size of a body. They helped explain how bodies formed aggregates or how smaller parts of matter accumulated into larger stuff, but the question of why a certain clump of atoms should become this or that specific thing could not be left to chance. There needed to be, Webster wrote, something else affecting inert matter that could guide it in a particular direction. He called that something, following van Helmont, a "seminary principle" or "active power," and gave to it the ability "to turn the matter aggregated into the nature of this or that stone."

Another example of the continuation of earlier influences into the age of "new learning" is the work of a Helmontian atomist named Thomas Sherley. For Sherley, van Helmont's elementary water was "a fluid body, consisting of very minute particles, and variously figur'd atoms or corpuscles." The accumulated mass was, however, full of pores that allowed for the penetration of "seminal beings" that thereafter directed particulate motions. From this "moving of matter," Sherley argued, "all the visible and tangible bodies of the world have their result" (Sherley, 1672; rept. 1978: preface). Key to Sherley's understanding, and something that has lots in common with earlier Paracelsian views, is the notion that the "seminal beings" contained in them "not only an Idea of the thing to be made, but also a power to move matter after a peculiar manner, by which means it reduces it to a form like itself." Sounds both mystical and mechanical all at once, doesn't it? And this is exactly the point. You didn't have to be Cartesian or Paracelsian, or to decide absolutely between alchemy and chemistry, in the seventeenth century. It was possible to combine traditions of several sorts. "I am," Sherley proclaimed, "no enemy to that rational way of explaining phenomena of nature used by atomical, Cartesian, or corpuscularian philosophers." These provided, he observed, very ingenious and true accounts of the way matter was put together and could be modified, often for the benefit of human beings. But such a materialist and mechanical approach to nature could not explain everything, and especially avoided the question of why matter took the form it did, and why it exhibited certain properties. If only the Cartesian philosophers would add to their reasoning "the powerful efficacy of seeds upon matter . . . we might," Sherley advised, "then hope to receive some satisfactory account of the generation of natural bodies" (pp. 123–124).

✳ At this point we need to take a closer look at Boyle's major text, his *Sceptical Chemist,* and describe in more detail both its essential purpose and its place in regard to the relationship between alchemy, chemistry, and the Scientific Revolution. Boyle's main aim in writing the text was to attack the claims of those who, without experimental justification or obvious demonstration, assumed the existence of principles and elements of various sorts in fashioning their chemical philosophies. Like others before him, Boyle was certain that part of the problem in giving certain substances the status of elements or principles had to do with a confusion of names and the use of "ambiguous expressions" that had come about due to the "phraseology of each particular chemist." "For I find," he writes, "that even eminent writers (such as Raymund Lully, Paracelsus, and others) do so abuse the terms they employ, that as they will now and then give diverse things, one name; so they will often times give one thing many names" (Boyle, 1661; rept. 1911: 113).

Boyle's arguments were especially aimed at the four elements of Aristotle and the three principles (Sulphur, Salt, and Mercury) of the Paracelsians. Anyone could claim to have resolved bodies into sulphur, salt, and mercury, but what types of substances were these? Did "sulphur" mean the marketplace stuff, or was it a reference to a kind of combustible principle? Moreover, he noted, there was no real agreement among Paracelsian "chemists" as to which properties these principles were responsible for in mixed bodies. "I could easily prosecute the imperfections of the vulgar chymists philosophy," says Boyle, "and shew you, that by going about to explicate by their three principles . . . all the abstruse properties of mixed bodies [and] even such obvious and more familiar phenomena as fluidity and firmness . . . chymists will be much more likely to discredit themselves and their hypothesis, than satisfy an intelligent inquirer after truth" (pp. 163–164).

For Boyle chemistry needed to become more philosophical without assumptions being made about the organization and fundamental principles of matter. Chemistry, he thought, should be raised up from a purely practical discipline to the status of a collaborator in natural philosophy that, by means of experiment, could penetrate into the actual design and configuration of bodies (Clericuzio, 1995). Boyle's message was clear, but it was not the first time that someone had viewed chemistry as a way of enlightening natural philosophy. Libavius, Brendel, and others, as we have seen, had already forced the doors of academic Aristotelianism and *chymia* had, by the time Boyle was writing, already gotten more than a small part of itself across the academic threshold. What Boyle meant by "real philosophy" was defined as "corpuscularian, atomical, Cartesian, mechanical," and within this definition chemistry was no longer an intruder at the table of philosophical discussion, but an invited guest.

Boyle was a believer in the corpuscular, or atomist, philosophy or, better said, he was a believer in a God who created a universe whose basic matter was composed of tiny bits, each of which followed His divine decrees. "We may without absurdity," he wrote in 1663, "conceive that God . . . having resolved, before the creation, to make such a world as this . . . did divide . . . that matter which he had provided into an innumerable multitude of variously figured Corpuscles, and . . . put them into such motions that by the assistance of his ordinary preserving concourse, the phenomena, which he intended should appear in the universe, must as orderly follow" (Boyle, 1664: 69). It was God who preserved motion and guided the physical actions of bodies. The case was similar, he observed, to the famous clock at Strassburg whose parts "are so framed and adapted, and are put into such motion, that though the numerous wheels, and other parts of it, move several wayes, . . . the various motions of the wheels, and other parts concur to exhibit the phenomena designed by the artificer." You needed an intelligent creator to "dispose of that chaos or confused heap of numberless atoms" brought into the world, and then "to establish the universal and conspiring harmonie of things" (p. 85).

Differences between substances were, in Boyle's view, due solely to the sizes, configurations, and motions of a body's constituent particles. Indeed, he wondered whether nature possessed any physical matter that actually deserved the name principle or element, or whether all substances were really different configurations of a particulate common matter. At the very end of his *Sceptical Chemist*, he notes that because "the violence of the fire" does not divide bodies into elementary substances, but rather makes new compounds out of them, "I see not why we must believe that there are any primogeneal and simple bodies, of which, as of pre-existent elements, nature is obliged to compound all others." Nor was there any reason not to think that nature simply altered or rearranged her minute parts when producing mixed bodies (Boyle, 1661; rept. 1911: 224). In a nutshell, Boyle writes, "as the difference of bodies may depend merely upon the schemes [arrangements] where into their common matter is put . . . the fire and the other agents, . . . partly by altering the shape or bigness of the constituent corpuscles of a body, partly by driving away some of them, partly by blending others with them, and partly by some new manner of connecting them, may give the whole portion of matter a new texture . . . and thereby make it deserve a new and distinct name" (p. 223).

For Boyle, then, Aristotelian elements and Paracelsian principles were out. What replaced them was a particulate view of matter in which all the tiny bits obeyed physical laws determined and sustained by God. For the most part, the chemical philosophy of Paracelsus had been ushered out the door. But what about alchemy? Did Boyle's experimental chemistry erase the alchemical tradition as well? Recently the historian of science Lawrence Principe has argued that *The Sceptical Chemist*, while clearly condemning Paracelsian chemists, nevertheless contained nothing that would justify viewing the book as anti-alchemical. In fact, Boyle himself, Principe notes, is an excellent example of the continuity of alchemical and chemical traditions during the age of the Scientific Revolution. Even his view of matter as made up of tiny corpuscles was not so much subversive to traditional thinking as tied, in many respects, to a corpuscularian tradition in alchemy stemming from the speculations of the medieval author Geber (Jabir ibn Hayyan) (Newman, 1994; Principe, 1994). In Boyle's opinion, there was no great distance to be crossed between admitting alchemical reasoning about transmutation and treating nature as a mechanical structure. One did not need to replace the other.

That Boyle accepted the reality of transmutation and the validity of claims about the powers of the Philosophers' Stone is clear from an unpublished *Dialogue on the Transmutation of Metals*, discussed by Principe and others. There, opponents of transmutation were soundly refuted with the report of an "anti-elixir" that, when projected onto molten gold, transmuted the gold into a base metal—an alchemist's nightmare, perhaps, but transmutation nevertheless. Although the *Dialogue* in its entirety never saw printer's ink, Boyle did publish its last section concerning the anti-elixir anonymously in 1678 as *An Historical Account of a Degradation of Gold* (Ihde, 1964; Principe, 1994). This was the real-life Robert Boyle, it has been argued. Only later, after his death, did Boyle's alchemy, his providential beliefs, and his uncertainties relating to the possession of natural knowledge become so embarrassing to advocates of modern science that his ideas needed to be culturally pruned and he himself transmuted into an exemplar of mechanistic virtue (Clericuzio, 1990).

Most of the alchemical tracts known to have been in Boyle's possession were contributions from a circle of friends and acquaintances. Sometimes he sought their help directly, however, and Principe has argued that Boyle's famous paper in the *Philosophical Transactions* (the main publication of the London Royal Society) on an "incalescent" mercury (a mercury that grew increasingly hot) was in fact a plea for help from alchemical adepts who knew the proper procedure for using mercury to produce the Philosophers' Stone. His own assistant was given the job at one point of overseeing the labors of a German alchemist brought to England and supported at Boyle's expense. Boyle himself became entwined within a "company" of alchemical practitioners seeking out ways to produce transmutations. An important figure within that circle was, as we have noted, George Starkey (also known as Eirenaeus Philalethes), a former member of Harvard College whose notebooks have a lot to say about quantitative techniques used in alchemical procedures at the time when Boyle was busy in the alchemical lab, and about the continued influence of Helmontian recipes in alchemical practice (Newman and Principe, 2002). Among Boyle's papers are hundreds of pages of laboratory processes, many relating to metallic transmutations and largely written in code. There is even a precise account of what took place on one occasion of metallic transmutation when he himself was a direct witness. Seeing was believing, and Boyle had no doubt of what he saw (Principe, 1998: 93–134).

So, what was Boyle after when he studied transmutational alchemy? On the one hand, it is clear that he actually hoped to create physical changes in bodies by means of preparing a Philosophers' Stone. On the other, Boyle's studies provided him with weighty evidence in defense of orthodox Christianity. Indeed the Philosophers' Stone, he believed, could also attract spirits and angels by means of what he called "conguities" or "magnetisms" (Principe, 2000: 215). There was nothing entirely new in this. Another Englishman, John Dee, had earlier mixed angel conversations with natural philosophy and sought procedural information in making the Philosophers' Stone through angelic contact (Harkness, 1999). Boyle, however, was not seeking angelic advice. What he wanted was to demonstrate the existence of God by actually producing the means, the Philosophers' Stone, to make God's spirits manifest.

✳ The skeptical empiricism of Bacon and Boyle, and the habit of making knowledge through experiment, stimulated natural philosophers, including alchemists and chemists, toward new discoveries throughout the seventeenth century. In Germany Johann Kunckel von Lowenstern amazed and confounded onlookers with the discovery of substances like phosphorus that exhibited curious properties. At the courts of the princes of Saxony and Brandenburg, the alchemist Johann Friedrich Böttger and the mathematician Count Ehrenfried Walther von Tschirnhaus collaborated in projects of fusing minerals in pursuit of another sort of *arcanum,* the secret formula for producing hard-paste porcelain. Böttger had proclaimed a knowledge of metallic transmutation in both Prussia and Saxony and the princes of both realms had sought to imprison him until he had made good on his claims and made gold. The collaboration with Tschirnhaus, however, resulted in something even more remarkable. Böttger's alchemical knowledge of how various stones sinter and melt at high temperatures mixed with Tschirnhaus's designs for building kilns that used burning lenses to concentrate solar heat. The consequence was the discovery of the white, translucent material that led to the establishment of Europe's first hard-paste porcelain factory at Dresden in 1710–not the Philosophers' Stone exactly, but, from the point of view of political economy, every bit as valuable. In fact, the awareness that industry and exports offered the best means to increase the wealth of territorial treasuries had already led to a merging of projects relating chemistry and commerce at the German court in Bavaria. There the central figure was a devoutly religious physician and court mathematician named Johann Joachim Becher (1635–1682) (Frühsorge and Strasser, 1993; Smith, 1994).

In Becher's view, no one—not Aristotle, not Paracelsus, not van Helmont, nor even Boyle—had yet got it right when it came to explaining the basic elements of matter. Especially he rejected van Helmont's view that the growth of plants confirmed the elementary nature of water. The growth of the willow tree in van Helmont's experiment, he argued, was not the result of water being turned into vegetable matter, but rather the result of earth being brought into the substance of the tree by means of water. In Becher's opinion, Earth and Water first separated from an original chaos, and their combination thereafter accounted for material existence. Different types of elementary Water and three different types of Earth brought about substances of various sorts. Three Earths, called oily, fluid, and vitreous, were especially responsible for the formation of subterranean things like minerals and metals, and some commentators have noticed that they bear a striking similarity to the three principles of Paracelsus, although Becher disapproved of their Paracelsian names. In his best-known text, called *Physica subterranea* (1669), a text that continued to expand with the addition of three supplements in the following decade or so, Becher wrote that "the first principle of minerals is a stone in fusion or an earth which is rightly called salt; the second principle in minerals is a fatty earth improperly called 'sulfur'; the third principle of minerals is a fluid earth improperly called 'mercury'" (quoted in Metzger, 1937; rept. 1991: 38). It was Becher's fatty or oily Earth, *terra pinguis,* that especially linked his ideas to an alchemical tradition that had long viewed the cause of combustion, sometimes called "phlogiston" (a Greek work that simply means inflammable), to be a principle of all bodies that would burn.

Like others before him, Becher declared that students of chemistry must be skilled in both natural philosophy and in the techniques of the laboratory. When studying nature, they must inspect the subterranean laboratory. But when making things—medicines and other useful compounds, for instance—their attention had to be geared to the laboratory above the earth, to the "superterranean" laboratory (Debus, 1977; rept. 2002: 458–463). There, one no longer simply observed nature, but acted on her so as to produce things that could improve the conditions of life. Chemistry was truly a public calling that required the practice of civic virtue. It led to industrial and commercial ventures and created wealth for one's prince and his people. While never doubting the truth of metallic transmutation, and finding no real problem in selling a process for transmuting silver into gold to the Dutch city of Haarlem, Becher nevertheless condemned those who pursued alchemical schemes as a means to private wealth. The preface to a later edition of his *Physica subterranea* underscored the opinion that nothing was more pleasing to such selfish and solitary creatures than to be extremely dirty, to be regarded poorly by the world, to squander their money and their reputations, and to turn themselves pale with drugs and poisons. Private alchemists had become part of a counterculture, in no way of service to anyone but themselves. "They live in coals, pollution, soot, and ovens and prefer these," the preface continues, "to the splendor of the court, economic and domestic order, public opinion, and the vigor of health" (Becher, 1703: preface).

✳ One of Becher's greatest admirers was another German chemist named Georg Ernst Stahl. Like Becher he accepted a close relationship between nature and art, endorsing the view that, through human industry, the material transformations accomplished in the natural world could be imitated for utilitarian purposes. In 1730, an English translation of one of Stahl's major texts, called the *Philosophical Principles of Universal Chemistry,* informed English readers that "the chemical and physical operations of Art and Nature differ as to time and place. Nature produces where it finds the principles; but the chemist collects his principles, and produces where he pleases: Nature produces when the principles meet one another, as it were by accident, but the chemist brings these principles together, at the time he would produce [in other words, at any time he wants]" (Shaw, 1730: 9).

Stahl's reflections on the relationship between nature and art point to yet another important connection between alchemy or chemistry and the experimental approach to natural knowledge. The historian William Newman has noted that it was primarily due to Aristotle that an unbridgeable gulf had long existed between things produced by art and those produced by nature. No artificial thing, in other words, could give insight into the structure of nature. Natural things, Aristotle maintained, had an innate principle of movement or change. The artificial product, on the other hand, was static with no intrinsic principle of motion or development (Newman, 1998: 11). A famous line comes from Aristotle's *Physics*. Because only nature possesses an inborn quality of motion, "men," he says, "propagate men, but bedsteads do not propagate bedsteads." The distinction between art and nature kept practical experience separate from natural knowledge until, it has been argued, Francis Bacon began to describe nature *as* art and thus allowed for natural knowledge to be gained through study of the artificial, or, as it has more recently been called, through "contrived experience" (in other words, experiment) (Daston, 1988; Dear, 1995). The point that Newman wants us to pay attention to, however, is that alchemists had been insisting for centuries that art not only uncovered the principles of nature by means of manifest tests, but that it could surpass nature as well. As with the Baconian notion of the "constraint of nature," alchemy and, later, chemistry had already made part of daily practice what later became a "Baconian" idea, manipulating and artistically recombining the particulars of mixed substances to draw forth a knowledge of their qualities and constitutions (Newman, 1998).

"Chemistry," Stahl writes, "is without contradiction one of the most useful arts, and it would be no exaggeration to call it the mother or instructress of other arts . . . she alone teaches us the work of God" (Debus, 1977; rept. 2002: 464–467). With the distinction between art and nature removed, the work of God was also the work of human hands. Producing natural knowledge proceeded by learning manual operations. One would not get far in understanding the operations of nature, Stahl declared, without the work of an "efficient cause," or an operator who makes change happen—a kind of artist. In chemistry, the "efficient cause" was the chemist (Shaw, 1730: 1–2). Stahl also knew that the chemist was the "efficient cause" of social and political well-being and was essential to the economic success of emerging nation states. Just as alchemical gold-making had often been tied to the personal ambitions of Renaissance princes, chemistry, he believed, now nurtured the competitive interests of increasingly centralized and bureaucratic political systems.

In the general sense, chemistry was "the art of resolving mixed, compound, or aggregate bodies into their principles," and then "of composing such bodies from those [same] principles." So, the object of chemistry was "resolution and combination," or, if it made more sense, "destruction and generation." What one ended up with was a theoretical understanding of the structure of nature, and, just as importantly, pharmaceutical, mechanical, economical, and practical know how.

To get started in the subject, Stahl believed one first had to understand that all bodies were either simple or compounded. Simple bodies were really basic principles into which all the compounded bodies could be reduced. In general, Stahl agreed with Becher, his favorite author, that the "immediate material principles of mixts" were Water and Earth, and that Earth was of three kinds depending upon the purposes they served: vitreous earth accounted for fusibility, oily earth for inflammability, and fluid earth for the mercurial nature of metals (Shaw, 1730: 3–8). Compound bodies were also of three sorts, which he called mixed, compounded, or aggregates. Regardless of their composition, however, all compounds, Stahl believed, were made of atoms. Nothing could be said of the shapes of atoms, as certain mechanical philosophers believed, because they were so small as to be indiscernible; but one could get a sense of what sorts of atoms there were by noting differences in their affinity for one another (in other words, their inclination to join one another in groups). Some materials, or atoms, liked to join with certain other materials, or atoms, and this readiness to join in compounds Stahl called "contiguity." So, while the properties of atoms could not be explained by referring to certain shapes, as Cartesians especially were in the habit of doing, one could get a sense of their qualities by recording the specific properties brought to a substance in the process of making a compound. All atoms, Stahl concluded, acted in similar mechanical ways, but various kinds of atoms possessed specific properties peculiar to them, and the only way to learn about which atoms possessed which properties was by making things—by forcing nature through art, and with one's own hands, to combine and dissolve, and then by comparing the resulting gain or loss of properties.

The extent to which Stahl might have been attracted to alchemical transmutations may never be known for sure. Some writers absolutely refuse to think that there was any affinity between Stahl and alchemy at all. And yet, it is clear that whether or not he believed that transmutations were possible, Stahl was altogether informed about current transmutational theory and practice, and he may have been attracted to some of it. Many of his remarks follow from the comments of his favorite author, Johann Becher; but some are also linked to the work of a French alchemist, well known in the seventeenth century, called Gaston Claveus. In fact, it is primarily from Claveus that Stahl in his writings records an alchemical process involving the combination of "philosophical Mercury" and "philosophical Gold."

Claveus had noted that "if an equal quantity, or less, of philosophical Mercury be mixed with philosophical Gold, and they are digested or cemented together . . . the philosophical Gold will perfect more or less of the Mercury"; and this, Stahl comments thereafter, "seems not improbable." The problem was how to make "philosophical Gold." One way was to produce it from common gold, but Stahl records another method in which vitriol (sulfates of iron or copper usually) were used. The reason given for the use of vitriol, he writes, "is that the substance which will combine with Mercury is probably an extract from iron or copper, and vitriol is nothing but iron or copper very subtly dissolved, and as the philosophical Gold is allowed to lie concealed in iron or copper, it must of necessity also lie concealed in vitriol" (p. 408). In regard to this process, however, one that had received the attention also of Becher, Stahl remained unconvinced. If anything were to be achieved in the transmutational art, it certainly would have to follow the principles agreed on by all the philosophers—in other words, "that the nearer the materials chose for their grand work actually approach the metallic nature, the better the operation will succeed." The most convenient method of all, he observed, would require the "animation of mercury with gold and silver" and the "philosophical calcination of Gold" before joining it with Mercury. However, whether any of this was truly within the reach of the chemist, Stahl does not tell us. These were musings, not laboratory exhibitions. For anything to claim a place in authentic chemistry, Stahl had a simple rule. "Its scientifical experiments must be well understood, and its observations personally viewed and manually performed."

One thing "personally viewed and manually performed" by Stahl had to do with the appearance of an ash or powder on a piece of tin when the tin was heated over a fire. As we have seen, the observation was not new. Metallurgists since the time of Biringuccio had noticed the same thing when pursuing their craft. When Stahl heated the ashes by themselves in a container nothing further happened to them. However, he noticed that something very interesting occurred if, when the ashes were still on the surface of the hot, melted tin, he added oil, tar, resin, or some other combustible fatty substance, and stirred the mixture. In this instance, the ashes themselves melted and reunited entirely with the original metal. The calcination of metals, Stahl argued, was a certain kind of combustion where the metallic calx could be thought of as a kind of metallic ash. The astonishing thing was that this ash could be transmuted back into the original metal when, he theorized, the "principle of combustibility" was returned into it as a result of being burned in the presence of certain substances. What it received back Stahl called "phlogiston"—a renaming of Becher's second, oily or inflammable earth. That which caused combustion then was, according to Stahl, an actual substance, that is, real, genuine matter. Everything that burned contained this stuff called phlogiston, and calcination was a kind of combustion, and so was life itself. Plants, for instance, lived on phlogiston that they got from the air and that became incorporated thereafter in animals and minerals.

✳ For Stahl, mechanical doctrines could explain a lot, but they were, he believed, incapable of untangling the processes of life. Both the living and the non-living were composed of particulate matter, and while what held matter together could be described in mechanical terms, what maintained life, he believed, could not. That which supported life and resisted corruption and decay Stahl attributed to the existence of an immaterial vital principle that he called the *anima,* or soul. The *anima* directed the activities of the body by knowing what the goal or purpose of each part of the living thing should be, and by then reifying itself in the material realm by becoming the directed motions of the individual parts. So, the vital principle affected living things by exerting its effect on the body through motion. Motion, however, was not life itself, only its "instrumental cause." Motion, in other words, was not an attribute of matter as the Cartesians and other mechanists assumed. Motion came from outside matter altogether as a kind of congealed presence of a rational and ordering universal soul. Motion was the way that the immaterial *anima* influenced and directed physical bodies. The entire outlook came to be called animism, and Stahl is one of its best seventeenth-century representatives.

In the human body, the *anima* forms a link between the mind and the body's physical parts, so much so that the body's illnesses can be ascribed to what we would call today psychic disturbances and emotional stress. The *anima,* he reasoned, perceived emotions and transferred its immaterial psychic constitution, by means of congealing itself into motion, to certain physical parts of the body. In a pregnant woman maternal emotions and anxieties, he believed, were transferred in the same way to the fetus.

Living things, because they depended on the functioning of parts that had specific intents and purposes, could never be reduced purely to chance combinations of particles and random mechanical movements. Some directive agency had also to be involved, and we have already noticed that such an agent had long before received a variety of names. The ancient physician Galen spoke of "natural faculties" that guided the specific function of each organ of the body. The Paracelsians spoke of an *archeus.* Van Helmont thought in terms of "seeds"; Boyle thought it was providence. Stahl advanced another, similar idea. In this description, an *ens activum*— an active, or vital, principle, operating through motion—ordered bodily structures and ensured their proper functions. The active or vital principle conveyed properties and qualities and bridged the worlds of mind and matter. It had no body, but was still biological. Mechanists cried foul!

CHAPTER SIX

THE REALITY OF RELATIONSHIP

In any discussion of the metaphor of the Scientific Revolution, the debate that usually takes place centers on the relative contributions of individual disciplines like mathematics, physics, chemistry, and medicine. Scholars write with great emotion, arguing whether the Scientific Revolution was an event that primarily had to do with the mathematical and physical sciences or whether it possessed chemical, pharmaceutical, or medical features as well. There is, however, another impression of this historical metaphor that you get when you look for its identity not *in* certain disciplines, but *between* them, allowing them all, to a greater or lesser extent, to push and pull on one another in the process of offering new interpretations of nature. Lines separating theoretical convictions were, during the period of the Scientific Revolution, far from distinct. Margaret Jacob, in discussing the cosmopolitan nature of early modern science, notes that the fuzziness of learned frontiers increased the possibility of social interaction between representatives of ostensibly different intellectual points of view and had a special consequence for the role of alchemy in the new science (Jacob, forthcoming: chap. 2). Alchemists became involved in experimentation, crossed national borders, and alternated between academies and courts. They also took part, as we have seen, in discussions of the mechanical structure of nature while, at the same, balancing those discussions with arguments for the presence of vitalist principles.

Picasso noted once, "This picture is not thought out and determined beforehand, rather while it is being made it follows the mobility of thought" (Ghiselin, 1954: 49). I think we have much to gain if we also follow the "mobility of thought" when imagining the Scientific Revolution. Doing so lets us think in terms not necessarily of either/or but of both/and. It allows supposed opposites like mechanism and vitalism, alchemy and physics, to coexist more naturally, and it offers a way for us to consider further how a process of making things can share a role in the process of creating scientific knowledge.

✳ At the same time Nicholas Lemery was publishing his famous *Course of Chemistry*, Friedrich Hoffmann (1660–1742) was finishing up his medical degree at the University of Jena. He left first for Holland and then moved to England, seeking out the already-famous Robert Boyle. Later, once settled again in German-speaking territory, a successful medical practice led finally to his selection as professor of medicine at the University of Halle, where Georg Ernst Stahl was also a member of the medical faculty. Like Lemery and others, Hoffmann has been claimed by historians eager for him to represent specific traditions. The label that most like to use to define his view of nature and the body is "iatromechanical," a combination of medicine and the mechanical philosophy, and a term whose precise definition varies from one commentator to another. It is, however, this very ambiguity and the eclectic nature of Hoffmann's approach to natural and medical knowledge that give his works and ideas special significance when we look for instances of the "mobility of thought" that characterize much of the Scientific Revolution.

Like Stahl, Hoffmann sought a guiding force for the movements that accounted for the activity of the body (King, 1964). Unlike

Stahl, however, Hoffmann conceived of this guiding force as an altogether material substance. It was, in fact, an ether, a component of the air and derived initially from the sun, that was, Hoffmann said, responsible for guiding bodily processes such as the movements of the blood and muscles, and accounted for the way the nerves functioned to represent the senses. Although significant parts of his natural philosophy have been linked to several earlier contemporaries who also held that ether was a universal principle of motion and natural change, it is Descartes whose views clearly stand out in Hoffmann's most important text, the *Foundations of Medicine* (1695). Unlike the notion of seeds or *anima,* the ether, in Hoffmann's view, conveyed a material order, as opposed to a psychic or spiritual design, to undifferentiated matter. By this means ambiguous and formless stuff became a specific thing. Moreover, in his account, both the world of nature and the smaller world of the human body traced their first principles solely to the two bulwarks of mechanical philosophy, namely, matter and motion.

All change in the universe is due, Hoffmann writes, to motion whose cause is God, "the greatest and best mechanic," who maintains all bodies in the universe according to their equilibrium, weight, measure, and arrangement. That which organizes the world is not something therefore that is metaphysically remote or spiritually inaccessible, but something close at hand—matter and motion. That means that the stuff of nature is also the stuff of art and the artist can artificially imitate and even improve on the material world. "And thus the artist who is skilled in the properties of matter, the laws of motion, and in precise calculations knows how to change bodies as he wills, to destroy them or put them together. In like manner, the physician, provided with these same principles can distinguish himself outstandingly by changing, resolving, and altering bodies in various ways, as in the example of well cultivated and diligent chemistry itself" (Hoffmann, 1695: 2).

The chemist, by forcibly rearranging parts, increased or diminished the possibilities of relationship between them, and this "changing, resolving, and altering" of bodies allowed natural philosophy to peer into "the recesses of nature and examine the hidden structures, proportions and mixtures of things." Experience could be manufactured, not just tolerated, and the type of experience that came about by compelling or coercing nature Hoffmann called "the first parent of truth." One might rest content with passive observations alone, but this provided only refinements in viewing nature from afar and did not necessarily lead to a perception of her secret facets. Preferable to simple observation were the techniques of those "who cultivate more deeply and precisely the study of nature, calling for aid on various experiments drawn from mechanics, anatomy, and chemistry."

Chemistry and mechanics also served a didactic purpose in medicine, and Hoffmann believed that the basic principles of medicine could be reduced into a "brief system . . . arranged by the easiest method, according to the precepts of sound modern mechanical-chemical philosophy. From this the whole science of medicine may be properly acquired in a short time." Disputes, controversies and "mischievous trivial questions" only got in the way. Thus Hoffmann ignored "any nauseous collection of opinions with which others (teachers) customarily weigh down and blunt the mind of the student." Instead, he notes, "I have provided what is true, what can be demonstrated and is supported by the principles of physics and mechanics." What was true was what was mechanical, and what was mechanical in medicine as well as in natural philosophy were the particles of Descartes and the material guiding force of ether (Hoffmann, 1685: 1–4).

Life itself was, in this view, not dependent on the presence of vital or spiritual principles, but was simply the result of another way of artistically arranging matter. Boyle had already referred to the body as a "hydraulico-pneumatical engine," and he had rejected references to "world soul" and "natural faculties" as the chaperons of generation, assimilation, and growth. And yet for Boyle and others, there was still an acknowledgment of an "untaught skill" that allowed parts of the body to act in pursuit of certain ends. Where did the design or idea for such action come from? From the structure itself? From God? Boyle, as we have seen, settled on divine wisdom and power as that which moved a passive mechanical "nature" according to its own ends. Not only did God give motion to matter, but in the beginning, "he so guided the various motions of the parts of it, as to contrive them into the world he designed they should compose (furnished with the seminal principles and structures, or models of living creatures)" (Giglioni, 1995). Boyle's notion of nature, one shared essentially also by Hoffmann, is of a clockwork, or better, an automaton whose design is there, as the historian of medicine Guido Giglioni says, by "primal contrivance," a structure "consisting of innumerable relations among the parts and the whole, and among the parts and themselves" (p. 256). The important thing to notice is that the guiding principle that designates the purpose or function of each member of the body is actually found in the *physical relationships between* each individual particle and each larger component. What Hoffmann argued as relevant for the body, Newton would suggest was also true for the macrocosmos. The reality of its being was to be found in the relations, especially in the attractions and repulsions, between things.

For Hoffmann, a change in the relationship between parts, prompted itself by the universal principle of motion, gave rise to all organic processes. Every phenomenon in nature and in the body took place as a result of a change with respect to a preceding circumstance. So the body is in constant motion, separating particles from the vicinity of one another, bringing them into the vicinity of others, and causing reactions within neighboring environments. "All that is artificial," said Descartes, "is also natural"; and in explaining what is natural to the body by artificial, mechanical means, Hoffmann held a great deal in common with other practitioners

of mechanical medicine, especially the Italian mechanists Marcello
Malpighi (1628–1694), Giovanni Alfonso Borelli (1608–1679),
Lorenzo Bellini (1643–1704), and Georgio Baglivi (1668–1707).
Malpighi's formulation of the body as a machine made up of
smaller machines is famous. "Nature," he wrote, "in order to carry
out the marvelous operations [that occur] in animals and plants
has been pleased to construct their organized bodies with a very
large number of machines, which are of necessity made up of extremely minute parts . . . Nature's method, then, . . . is to make use
of little parts, such as salt, filaments, and the like, and with these
minute things to construct every work . . . Just as Nature deserves
praise and admiration for making machines so small, so too
the physician who observes them . . . must also correct and repair
these machines as well as he can every time they get out of order"
(quoted in Giglioni, 1997: 156–157). That physician, then, is a mechanic who understands the intricate relationship of the body's
parts and the chemical processes that make them work.

Chemistry, as well as medicine, was physics, and while Hoffmann
and others gave credit to Paracelsus and van Helmont for introducing chemical remedies into pharmacy, they also held them account-
able for introducing an intellectual heritage damaging to medicine.
Notions like seminal ideas and active powers were not part of the
true rationality of medicine, a rationality that began not in theoretical speculation but in the experience of treating patients and that
demonstrated its conjectures through the practical ability to restore
health. Indeed, Hoffmann, it has been claimed, viewed himself not
primarily as a philosopher or scholar but as a practicing physician
for whom knowledge came to light by means of treating concrete
individual cases of illness. "All things in theory," he observed, "are
truly better distinguished at the bedside as they are conferred upon
health" (Müller, 1991). The greatest certainty in medical understanding arose as a result of comprehending the direct and immediate causes of all those things that could be observed in the body at

times of sickness and well-being. These were nothing other than mechanical and chemical causes, and Hoffmann criticized van Helmont and other "chemists" who advocated a single remote cause, like the *archeus,* which eliminated any necessity for inquiring after more immediate and manifest reasons for the body's activity.

Both chemistry and mechanics were indispensable parts of rational inquiry and contributed to the perfection of medical science. Mechanics, however, was the maternal discipline, and Hoffmann argued that whatever could be claimed in chemistry needed to be derived from mechanics itself. In that way mechanics would offer useful explanations for natural appearances as well as for matters related to the human body, such as the dissolution processes in the stomach and intestines, and could shed light as well on the origin of illness. Paying attention to mechanics had led William Harvey to the discovery of the circulation of the blood, and other discoveries would certainly follow from the combination of chemical philosophy and experimental mechanics. The two systems worked together to provide a rational system that disclosed the operations of nature as well as the functioning of the body. Without them, Hoffmann proclaimed, natural philosophers and physicians alike rested their claims to knowledge on a chimera. Nor did combining mechanics with past traditions render any real insight. As he noted in a later, very imposing exposition of "rational medicine," some even in his own generation had framed their ideas partly from the corpuscular philosophy of the Cartesians, partly from the potent salts and sulphurs of the chemists, and partly from the schools of the metaphysicians; but these had only succeeded in complicating matters by supplementing obscurity without offering any help to the construction of solid theory and rational medicine (Hoffmann, 1738).

Traditions like the mechanical philosophy that we might view as having tidy boundaries were, in the world of early modern experience, far more unkempt, cluttered sometimes with bits and pieces surviving from other philosophies of nature. Hoffmann's devotion to mechanics is no exception. Although he has been categorized as an iatromechanist for purposes of historical representation, Hoffmann himself would not have recognized any such entity and, in fact, steadily refused to define his views in terms of any specific sect or hypothesis. A chapter from one of his better-known texts is called "Concerning Eclectic Medicine," and there he promises "to examine everything by its own consequences and to select those things which are of use and agree with [experiential] truth." Not surprisingly, the six copious volumes of his collected works have provided evidence of a variety of influences (cf. Rothschuh, 1976), and it is not out of place for us to take note of an interpretation of his thinking slightly at variance to strict mechanist descriptions. In fact, one historian who spent a long time examining Hoffmann's work concluded that the division between material and immaterial existence in Hoffmann's comprehensive natural philosophy was not absolute, and that his first principle of motion was actually akin to a soul or vital principle analogous to a "spirit endowed with mechanical powers" (quoted in King, 1969: 27). At bottom was the problem of how mind related to matter. What was it, after all, that accounted for the "more noble" powers of thinking and reasoning in human beings? As much for Hoffmann as for Descartes, mind was immaterial, and the distance between mind and body had to be bridged by a metaphysical "power" that, Hoffmann assumed, was somehow able to relate to, and influence, material particles. "In essence" one interpretation goes, he "took a sort of animistic view, but by verbal juggling *called* it mechanical" (King, 1970: 190–191).

That may be going too far. Hoffmann really liked mechanical descriptions and made no attempt to disguise the fact. On the other hand, it is true that some natural philosophers, such as the German philosopher Gottfried Wilhelm Leibniz (1646–1716), considered souls and bodies to be closely related and had no trouble in thinking of this relationship as part of the mechanical structure of nature. Leibniz explained: "I believe that everything in fact happens mechanically in nature and can be explained by efficient causes [motions], but at the same time everything also takes place morally, so to speak, and can be explained by final causes [the design or purpose of creation]. These two kingdoms, the moral one of minds and souls and the mechanical one of bodies, penetrate one another and agree perfectly on account of the Author of things, who is at the same time the first efficient cause and the final end" (quoted in Rutherford, 1995: 215–216).

As a physician Hoffmann was, of course, mostly concerned with the processes of life; and while grappling with the relationship between mind and body, he never expressed any doubt that the laws of mechanics could explain the functioning of organisms. To account for the properties of various parts of the body, he reduced them to the peculiar actions of chemical particles. Disease itself was simply a variant of particulate motion. Thus, physics and chemistry combined not only to describe how the body functioned, but why it operated as it did. The presence of certain particles and their motions caused other particles to act and react in certain ways. From the action and relationship of parts large and small, structured and organized, life emerged. It did not seem to matter if one agreed with Malpighi and Hoffmann that the body was a machine governed by smaller machines within it, or with Leibniz that the body was an organic creature composed of smaller organic creatures. In either case, what was really at issue was how those parts, mechanical or organic, related to one another so as to produce the living reality, the fact, of purposeful, animated being. Only God, the ultimate maker, comprehended the harmonious relationship of all the necessary parts. Yet it was clear that both form and function were the products of mechanical virtuosity. The design and intent of the body's parts emerged as a result of artistically arranged mechanical structures. The literary monster of Dr. Frankenstein may still have been several generations into the future, but both mechanists and vitalists would have agreed that just as nature reflected a certain

artistry, so also did life itself. In tracing the elements of the artistry
of life, chemistry stood at the heart of the matter.

✳ When things begin to look too clear historically, odds are
you are missing something. At the opening of his book of *Physico-
Chemical Observations,* in which he recorded intriguing experiments with, among other things, phosphorus and the action of
light on silver salts, Hoffmann denounced the perplexing and confusing terms and figurative use of language found in the texts of alchemists and chemists alike. Older writers like Paracelsus, Isaac
Holland, and Basil Valentine and more recent authors like Johann
Rudolf Glauber, Becher, and Kunckel were similarly castigated for
not sufficiently explicating experience and for not transferring
experience into practical use. Knowledge, Hoffmann admonished,
could not be gained by collecting the enigmatic opinions of solitary
alchemists and chemists (Hoffmann, 1722: preface). Truths, he was
sure, came about only through rational (mechanical) interpretations of firsthand observations. Any other approach was useless.
Hoffmann's position in regard to acquiring knowledge is definite
and distinct. It is also, from the point of view of clarifying the type
of knowledge that should count in the Scientific Revolution, just a
little awkward. The problem is that by rejecting the collection of
ancient and contemporary alchemical opinions as being in any way
appropriate to science, he dismissed an approach to comprehending the natural world that had been adopted by one of the most important figures in anyone's definition of the Scientific Revolution,
Isaac Newton (1642–1727).

Newton was indeed a great collector of alchemical wisdom in the
form of transcriptions, extracts, and collations of ancient, medieval,
and contemporary alchemical authorities. For years he labored over
the construction of a chemical index, an inventory of chemical and
alchemical writing arranged by topic that, in its final form, comprised a volume of more than a hundred pages with 879 different

THE REALITY OF RELATIONSHIP

headings. Another text of "Notable Opinions" consisted of quotations from seventy-five printed and handwritten alchemical sources (Westfall, 1980: 357ff; Principe, 2000: 204). The idea was to gather together as indices, collections of translations, and alchemical collations and compendia the various remaining bits and pieces of what Newton believed to be a once-coherent body of ancient alchemical wisdom that had become fragmented and jumbled up over time. Alchemical truths, Newton apparently thought, might thus be revealed as a result of comparing texts, names, and references and by looking for consensus and agreement among alchemical authorities.

Most of us know of Newton the mathematician and experimenter whose discoveries, dispersed in the two well-known texts the *Principia mathematica* (1687) and the *Opticks* (1704) altered the direction of thinking in natural philosophy and experimental science. Yet Newton was moved not only by deductive reasoning. He was also firmly committed to the belief, a very common belief connected to Renaissance traditions, in the existence of a *prisca sapientia* (a pure, ancient wisdom), that is, a unified body of pristine knowledge believed to have been bestowed on human beings by God at the outset of human existence. Human history, as a history of sin and corruption, was in part the history of the loss of this originally pure knowledge. God, apparently, only said things once, and if human beings chose to ignore the message, so much the worse for them. However, all was not lost. By bringing philological skills to bear on the analysis of ancient texts believed to be closer to the original revelation, Newton and others supposed that one could still catch a glimpse of the archetypal God-given truths that had been known in remotest antiquity.

Newton, for instance, was delighted, but not surprised, to find an inkling of his inverse square law of gravity—bodies attract one another with a force that is proportional to the product of their masses and inversely proportional to the squares of their distances apart—in ancient documents. There he found reference to the fact that "if two strings equal in thickness are stretched by weights appended, these strings will be in unison when the weights are reciprocally as the squares of the lengths of the strings." The ancients, Newton believed, knew all about the inverse square law, and had extended such a mathematical knowledge of musical harmonies to the problem of comprehending planetary motions. "For Pythagoras, as Macrobius avows, stretched the intestines of sheep or the sinews of oxen by attaching various weights, and from this learned the ratio of the celestial harmony . . . and consequently, by comparing those weights with the weights of the Planets and the lengths of the strings with the distances of the Planets, he understood by means of the harmony of the heavens that the weights of the Planets toward the Sun were reciprocally as the squares of the distances from the Sun" (quoted in McGuire and Rattansi, 1966: 116–117). Ancient texts were by no means worthless. Through them, Newton maintained, one distinguished an originally revealed knowledge that was now lost. There were to be no further revelations; but one could recover knowledge, nevertheless, by means of a different method— through empirical science and by the sweat of one's experimental and mathematical brow.

In his reading of ancient texts, Newton was especially fond of writings relating to the ancient Egyptian magus Hermes Trismegistus. Besides the comments of Hermes, however, he collected a great many other alchemical opinions, and some of these left their marks on his developing ideas about the construction of matter. In fact, several historians have noted a link between certain parts of Newton's understanding of nature and alchemical opinions expressed in texts to which he had access. Especially influential in this regard were the deductions of an accomplished laboratory adept that we have already met, George Starkey. Both Betty Jo Dobbs and more recently William Newman have, for instance, pointed to the concept of chemical mediation (the means by which two unsociable bodies are made sociable by means of a third) as having been received by Newton through the mediation of Starkey. Newton's notebooks show a familiarity with several procedures originally described by Boyle and Starkey (Dobbs, 1975: 220ff; Newman, 1994b: 229–239) and suggest also an alchemical heritage to Newton's belief in a universal matter composed of particles. The largest particles of every sort of matter, he theorized, were themselves made up of very subtle sulphurous or acid particles surrounded by larger earthy or mercurial particles, the latter piled up like rings or shells around the volatile center. Not only, then, does Newton revisit the sulphur-mercury principle of medieval alchemists in his description of matter, but every substance, he held, was composed of particles analogous to tiny universes with a "chaos" or "heaven" at its kernel and with a less subtle "earth" at the surface. Doesn't sound much like the Newton of physics textbooks, does it?

According to Betty Jo Dobbs, the publication of whose book *The Foundations of Newton's Alchemy* in 1975 represented a sea change in the perception of Newton's relevance to the Scientific Revolution, Newton knew a great deal about chemistry and could best be understood as a "scientific alchemist." Much of his intellectual life, she argued, especially that part of it after 1675, was given over to ever-renewed endeavors aimed at cementing together alchemy and the mechanical philosophy. The alchemy he studied, of course, was initially very much of the esoteric sort expressed in symbolic language. What Newton was clearly after in this type of reading was an indication, by means of interpreting that language, of a direct line of descent in alchemical knowledge stemming from the earliest, and therefore purest, ancient sources. The means of testing the assertions thought to originate in ancient wisdom could then be done experimentally, and there is good evidence to suggest that Newton believed some of these remaining alchemical fragments had indeed been confirmed by rigorous experimental analysis.

Several important alchemical concepts can also be detected in

Newton's writings. The first is the ancient idea of the existence of a "universal spirit" that gives rise to all the various sorts of material substance in the world. In Newton's hands this idea joined a mechanical system of particles in which particles of a certain intermediate size (supposedly derived from the universal spirit) acted, as already mentioned, as mediators bringing about a kind of congeniality between different, less companionable types of corpuscles. Most important, however, Newton also adopted the alchemical notion of active principles in nature that accounted for attractions and affinities between bodies. According to one interpretation, it was this originally alchemical notion of active principles operating within the interstices of very porous matter that formed the seedbed for a new concept of force capable of universal action—one that not only accounted for the powers operating in the terrestrial and celestial realms (the force of attraction that explained the fall of an apple and the motion of the moon, for instance), but for the powers operating *inside* matter that furnished the *internal* bonds between the particles that constituted material existence itself (Dobbs, 1975: 230–231).

Nevertheless, for all the attention to alchemical traditions and experimentation, Newton himself was not, as far as we know at present, actively involved in attempted transmutations. What mattered more was the role that alchemical conjectures played in a different sort of intellectual endeavor, namely, in proving the continuing existence of divine agency in every part of the physical world. On December 21, 1705, Newton's later biographer, David Gregory, recorded his friend's opinion about a question of great concern, especially to Cartesian mechanists. The question was, What, if anything, filled the space between objects in the heavens? Gregory wrote: "The plain truth is, that he [Newton] believes God to be omnipresent in the literal sense" (quoted in Dobbs, 1991: 191). God directly intervened in his own creation, according to Newton, and was literally present in and between all things. That which was called universal attraction was the physical action of God, the "cosmic mediator." Just as alchemical mediation made possible the fusing of disparate substances, this divine action was the secret reality behind the coherence and physical order of all matter, whether that matter was as small as a particle or as big as a planet. That same divine action revealed itself in magnetic as well as in electrical relationships, where it accounted for cohesion, attraction, and repulsion. Divine activity was certainly present in the vegetable spirit that, Newton believed, was responsible for generation and nutrition. But no matter in what form it was found, this absolute force operated on all passive matter. Whether one studied how substances combined and dissolved through chemistry or how, by means of mathematics and physics, the planets continued in their courses, the underlying reality was the same—the absolute force of attraction, the revealed dynamic presence of God.

As we have seen, the mechanical philosophy featured by Boyle and Hoffmann was fundamentally Cartesian. Only extended matter (that which could fill a volume) and motion were acceptable as its principles. Particles of bodies would adhere to one another, Boyle thought, because of their relative shapes. To think that certain particles could have some sort of an affinity to others was to attribute to them specific extra-mechanical properties and thus amounted to an invitation to rejoin the dance with occult qualities. It is now well established that a major influence in Newton's dissent from the Cartesian view was his conviction that treating bodies only as something filling a space led ultimately to atheism. At the same time, he was equally convinced that thinking of matter as possessing inherent (occult) qualities was to admit that the substances of mind and body were the same, and to imply another kind of heresy, namely that God was nature itself. The solution, however, was not to distance God from the material world, but to keep God's hand permanently connected to the actions of the physical universe. "The religious Newton was never at odds with the scientific

Newton; quite the reverse," says historian of science Margaret Jacob (Jacob, 1997: 65). While the origin of the concept of universal gravitation and the calculations that disclosed nature's design can be linked to the inventions of a brilliant mathematician, the genesis of the idea of universal attractive force and the wish to demonstrate the power of divine will in all matter probably had much to do with reading *Genesis* itself—and the writings of Hermes.

Newton would show that along with "brute and stupid" matter there existed bodiless realities like "mass" (no pure Cartesian, committed to a view of matter as simply something that filled a space, could have dreamt of this) and "gravity." This really real reality, a reality of relationships, was what held both the planets in the universe and the particles of matter together. It was a reality, however, that did not need particles or planets to exist, because it existed before them—and had always existed as an attribute of God. Such ideas are quite distant from the ones most people associate with Newton's scientific achievements. They are, however, part of a grounded historical reality in which Newton gets to tell us what was important to him in terms of the world in which he lived. Trying to make Newton fit the "logic-tight compartments" of modern science and to sculpt his relevance to the Scientific Revolution solely in terms of classical physics and celestial dynamics, is, in this regard, to misconstrue his own lived experience in which theology, alchemy, mathematics, and physics were all active parts of an intellectual universe.

✳ In a famous comment in *Opticks,* a book about the nature of light and colors, Newton proposed "to find in specific attractions the explanations for all the reactions studied in chemistry." The principle of attraction could enlighten chemistry, Newton thought, when chemical qualities themselves were treated as special instances of universal forces. In the early eighteenth century, the question of particulate attraction became the special focus of two of Newton's fellow countrymen, the mathematician John Keill (1691–1721) and a physician named John Freind (1675–1728). The subject of how the particles of matter attracted one another also crossed the channel to appear in Cartesian Paris in a carefully written report to the Parisian Academy of Science presented by Étienne Geoffroy (1672–1731) in 1718.

Freind especially took the principle of attraction to heart in a book called *Chemical Lectures* (1712), which was based on a set of lectures given at Oxford in 1704. The book began with nine postulates, the last of which was that "the force by which Particles cohere among themselves arises from [Newtonian] Attraction, and is chang'd many ways, according to the various quantity of Contact" (Freind, 1737: 10). For Freind, solidity or hardness was really not a thing in itself, or even a state of being, but a force whereby the particles of a body resisted separation. The resistance, he said, "arises from a mutual Cohesion of its Parts. And Cohesion is . . . always proportional to an Attraction that necessarily resides in all Matter." This attractive force, according to Freind, was strongest between particles at points of contact, and consequently bodies yielded more slowly to separation "in proportion to the number of Points they touch one another in." Simply said, the more points of contact between particles, the greater the power of attraction and cohesion. Spherical bodies touched one another only at one point. Thus their power of cohesion was relatively small. Their particles "easily give way to every little Shock, and are put into Motion, whether it be by Nature or Art, [and] there fluidity takes place" (pp. 17–18). If the force of cohesion were proportional to the quantity of matter, or to the weight of bodies, then one might be able to determine how much force was necessary to melt or to change the state of a substance by simply knowing its specific gravity. However, Freind continues, "because the same quantity of Matter may be so variously dispos'd [shaped] that in one Body there shall be a much greater Contact than in the other," simply relying on attraction alone was insufficient to estimate the force of cohesion between particles. The actual, physical shapes of particles were still important to consider in matters of chemical composition and dissociation. Nevertheless, while still acknowledging Cartesian shapes as important, Freind's observations led him to an important conclusion. In his judgment, chemistry was a matter of proportions (relative strengths of attraction between differently shaped particles) and that meant that doing chemistry had to involve doing mathematics.

Freind's book is in many ways a defense of mathematical reasoning in chemistry; and his greatest fear was that, in doing so, he would displease those "chemists" who preferred to trace non-quantitative principles of vitalism into the laws of nature. But things do not always turn out the way one expects; and it happened that the loudest critics of his text were not animists, but Cartesians, especially a reviewer who anonymously voiced the opinions of an entire group of German scholars in a scientific and philosophical journal called the Leibzig *Transactions*. The fascinating thing is that, in defending his text against German criticisms, Freind found it necessary also to defend "occult qualities" (in other words, mysterious immaterial powers) and did so, ironically enough, on the basis of mathematical demonstrations provided by Newton.

"The Grounds upon which I proceeded in my Theory of Chymistry," Freind proclaimed, "were the Principles and Method of Reasoning, introduc'd by the Incomparable Sir Isaac Newton; whose Conclusions in Philosophy are as Demonstrative, as his Discoveries are Surprising" (pp. 173–174). The method of the Cartesians, however, had been, he notes, "to assume an Hypothesis [the existence solely of matter and motion] which has no foundation anywhere, but in the imagination only; and in general terms, to tell us, how everything in Nature may be produc'd according to that Hypothesis, without being able to give a clear and satisfactory account of one single Appearance" (p. 175). Newton assumed, on the other hand, "nothing but Observations and Experiments, which are evident to the Sense of all Mankind." From these he deduced demonstrative conclusions that were able to explain many phenomena in nature. Yes, Freind admitted, the "universal Tendency of Matter to Matter," called attraction, was termed by some "an Occult Quality, and I believe it will always remain so," for not even the greatest philosopher could show "how it may be produced mechanically." And yet, no matter how mysterious it is as a cause, the force of attraction could not be called a mere figment or hypothesis "since the Existence of it is as undeniably prov'd, as that of the Sun or the Planets." Attraction was a principle of nature, an occult cause if you like, but it was nevertheless bonded to matter and, Freind wanted to know, "what Reason can there be, why we may not make use of it in Philosophy? And shew how it is the real and adequate Cause of a great many [other] Effects, which we daily observe" (p. 177).

Freind argued that the true way to proceed in philosophical inquiries was to discover the properties of bodies by means of experiments "and then, without any further Search into the Cause of such properties, (which perhaps are insearchable) to explain the particular Phenomena, which depend upon them" (p. 178). In this way Archimedes discovered the principles of mechanics and the laws of hydrostatics without determining the cause of gravity and fluidity. But the Cartesians would have to reject these discoveries "because they are founded upon such Properties of Bodies, as have unknown causes; and cannot be explained, without admitting Occult Qualities" (pp. 178–180). Attraction was not an "hypothesis" invented to solve other phenomena, but was itself a phenomenon in nature. Moreover, to hold, as did the disciples of Descartes, that everything "results from the Essence of Matter and the unalterable Laws of Motion," would be to take away the necessity of "a Supreme Infinite Intelligent Being, who Directs and Rules the Universe" and would serve only to "furnish the Atheists with Arguments to defend and support their Impious Cause" (p. 189).

If you haven't noticed, we are in the theater of guiding force once again, and the fantasized play we are watching has Newton in the title role with Freind the main supporting actor. Directions in our make-believe play call for Newton to be off stage and the action to be set in the land of chemistry. "What do you mean by insisting on a figment like attraction?" says a man we don't know, but who in the script (the Leipzig journal) is called "Mr. L." "Are we supposed to return to the old refuge of ignorance where sympathies, antipathies, and qualities reigned over reason?" "But attraction," says Freind (and I paraphrase), "although occult, is not like these— not dark, obscure, or simply fictitious, but demonstrable by means of mathematics, experiment, and observation. Moreover, don't the Cartesians have their own fictions such as vortices and subtle fluids?" And even Mr. L has to admit that there must be an active principle somewhere existing in nature, "for Bodies once put into Motion, and then left to themselves, will not [otherwise] produce such regular and constant Appearances, as we daily observe." Whatever that active principle is, Freind asserts, "it must at last be resolved into an Occult Quality; for as yet we are not able to find out any other cause for it [other] than the Will of an Omnipotent Being" (p. 190).

If among the cast of characters our imaginary play had included Friedrich Hoffmann, what Freind has to say next could well have been addressed to him. "Those indeed who pretend most to Mechanism, place this active Principle in the Aether, or some extremely subtil Fluid; but then I wou'd ask the Question, What is it, that actuates this Aether, and constantly preserves it in Motion? How comes it to pass, that contrary Motions do not destroy one another? And what is it, that determines these Motions, to produce such particular Effects, and no others? These must necessarily be Occult Qualities residing in the Aether" (pp. 190–191). What a clever ending. The Cartesians themselves can't get along without occult qualities even though they deny their existence. The curtain falls, and the audience, expected to be complicit in the plot, goes home saying, good show.

Someone who might have liked this little pretend drama, but who also had to know that it would never play well in Paris, was a celebrated Frenchman named Étienne Geoffroy (1672–1731). Geoffrey had been an apothecary before studying medicine and visiting England during the heyday of Newtonianism. Once back in France, he became professor of chemistry at the Royal Garden and professor of pharmacy and medicine at the College de France. Geoffroy knew Paris and, more importantly for our purposes, he knew the mind of the Parisian chemical and medical establishment—and that mind, in the second decade of the eighteenth century, was still in great part a Cartesian mind. Thus, in 1718 and again in 1720, when he presented a report concerning the actions of substances on one another demonstrated in terms of the affinity of one thing for another, he was careful not to use the word "attraction." Instead, on the basis of collected observations, he presented the French Academy with a *Table de rapports* (things sound so much better in French), that is, a list of the various degrees of "rapport" between different chemical substances. Geoffroy's list was ac tually a table of the intensities with which certain chemical substances liked to combine with other chemical substances. It was a table of relative "attractions" without the use of that term.

Metallic substances, acids, alkalis, sulphur, and resins, he was able to show, liked to combine with some substances in preference to others. Matter, it seemed, liked to commit to relationships, but only until something better came along. Thus Geoffroy was able to show that the strength of the affinity, rapport, or attraction of certain bodies for certain reagents varied from substance to substance. When two substances combined, the addition of a third, with more affinity for one of the substances than the other, caused a separation from the unpreferred substance in the compound. But what was this "rapport?" It seemed like the question had been asked before when Newton and his followers wondered what a body was doing when it "endeavored" to attract another. What kind of a physical reality was one talking about, something material or something not? Could nature be such that she described herself according to mechanical laws but still possessed principles or occult properties? And, in designing a corresponding and coherent natural philosophy that could explain such operations in nature, was there any affinity between the animist views of Stahl and the mechanist opinions of Hoffmann? Moreover, what would happen to that possible relationship with the entrance of a third, Newtonian view of nature?

One Frenchman, Jean Baptiste Sénac (1693–1770) attempted to discover a rapport between the principle of attraction in Newton and the vital principle of Stahl in a work published at Paris in 1723. For him, purely mechanistic explanations (or explanations based on the shapes of particles only) were insufficient to account for the various phenomena observed in chemical reactions. Instead, he invoked the (occult, but measurable) power of magnetism as that which brought together, for instance, the particles of gold and *aqua regia,* shown by experiment to be acutely attracted to one another. Purely mechanistic explanations for chemical effects were also thought inadequate further north, at Leiden. There, Hermann Boerhaave asked once again a very old question, one found already in Aristotle, and one of the central questions of alchemy. How do bodies become mixed?

Considering several kinds of solvents (also called menstruums), Boerhaave began to write about affinities. "We easily perceive," he explained, "that many [solvents] unite bodies together, as well as separate them into their minutest parts." It was, he noted, a common observation that when the particles of some solvents had dissolved their solvends, they then united themselves to the particles of the body dissolved and formed a new compound body, "oftentimes very different in nature from the simple, dissolved one." So, new bodies can be formed from the division, separation, and reuniting of different kinds of particles, and the process is begun with the help of a certain dissolving substance. "But this now becomes particularly remarkable," Boerhaave continues, "when only some of the particles of the solvent and solvend are united together into one mass, whilst, at the same time, others are not admitted to this new concretion, but appear in a different form." Some particles of the solvent can be induced to ally themselves with some of the particles of the body dissolved, and become chiefly united with them.

But what would cause the particles of the solvent to disengage from one another and to associate themselves rather with the particles of the solvend? Furthermore, when the particles of the body that was dissolved were separated from one another by the action of the solvent, why then would they combine with certain kinds of particles of the solvent rather than splitting off completely from any compound and forming their own homogeneous bodies (bodies made up of only their own kind of particles)? "This, Gentlemen," Boerhaave instructs, "I would desire you to take particular notice of; for it highly deserves your observation."

I'll say it deserved attention, and lots of people in the early eighteenth century were attending to it. When demonstrating this "affinity of nature" to students, Boerhaave, like Sénac, also reached for gold and *aqua regia*. This was a dramatic demonstration of the power of attraction. Even though the particles of gold were eighteen times heavier than the particles of *aqua regia,* they were so strongly united together that the gold remained suspended in the resulting yellow fluid. "Is it not plain, therefore," Boerhaave half asks and half tells his students, "that between every particle of the gold and Aqua Regia there is some reciprocal vertue, by which they attract, and come into a close union with one another." There was "a certain power" that allowed particles to "endeavor to associate" with the particles of another substance, and this endeavoring could not be explained simply by referring to matter and motion. Dissolution in one compound was caused, oddly enough, by a stronger attraction to particles of another substance, and this was no mere mechanical thing. "Here, therefore, we are not to conceive of any mechanical actions, violent propulsions, or natural disagreement, but there seems, on the contrary, to be a sociable attraction and tendency towards an intimate union" (Boerhaave, 1735: 390–391).

Boerhaave was certainly happy to use mechanical metaphors when discussing chemical reactions. In dissolutions particles acted like wedges, insinuating themselves between other particles and separating them. But he was also sensitive to the fact that relying solely on mechanical metaphors did not speak to the cause of such motions. In some reactions, however, like crystallization and certain precipitations, Boerhaave accepted that the cause in question was a special property characteristic of a specific body rather than one shared equally by all. In other words, there existed not just different affinities, but different rules by which affinities occurred. Even some French Newtonians objected to this. All reactions, they believed, could be deduced from the law of universal attraction.

✳ Like Geoffroy, who, at the same time he prepared his table of empirically derived affinities, thought that iron could be artificially created in the combustion of vegetable matter, Boerhaave also mixed mechanical and Newtonian thinking with older alchemical assumptions about nature. Metals, for instance, were for Boerhaave not simple structures but combinations of principles recognizable to any medieval alchemist. Gold, he noted, "consists of a most pure, simple matter, very like Mercury fastly held together by another exceeding subtil, pure, and simple principle, which being intimately dispersed through the whole, firmly unites the particles of the former both with itself, and with one another: these two principles are supposed to be Mercury and Sulphur" (p. 26).

By the early eighteenth century, then, chemistry had become a major part of the new, experimental science. And yet, many of its questions were still the questions of traditional alchemy. The "subtil, pure, and simple principle" that Boerhaave invoked as that which held the particles of gold together was another reference to the unseen organizing hand of nature, discussed by vitalists and mechanists alike, that wove elements or particles together and thus gave form and function to specific things. As we have seen, alchemists, physicians, and philosophers described this principle, this guiding and directing force of nature, in various ways—sometimes petitioning the hand of God, and sometimes by referring to smaller machines operating within larger ones. Sometimes, as with Newton, they did both. Newton's force of attraction, a mathematically demonstrable occult quality that represented the presence of God in nature, was, in this way, a new verse to an old song—a long-sung ballad that looked for a reality in the relationships between things as well as in things themselves. In whatever way Newton's understanding of the forces of nature may be received today, and however much Newton himself may be summoned forth as one of the paternal figures of the Scientific Revolution, it is important to remember that questions posed by alchemy, and his attempts to pursue answers to those questions by means of alchemy, helped him to be wakeful to hidden patterns in nature. Newton may still be regarded as a genius, even if part of his ingenuity was rendered in the service of alchemy. Looking at the world as Newton saw it, alchemical knowledge still promised, as it had for centuries, insights into the order of creation. Indeed, just as Newton helped to describe a world in which a certain physical reality existed in the relationships between natural objects, the reality of his own genius may well have had its origins in the intellectual relationships he pursued between mathematical, theological, and alchemical matters of inquiry.

CONCLUSION:
VARIETIES OF EXPERIENCE IN
READING THE BOOK OF NATURE

When Antoine Laurent Lavoisier (1743–1794) married Marie Anne Pierrette Paulze in 1771, he was slightly more than twice her age and she was not yet fourteen years old. Their marriage lasted to the time when Antoine died as a victim of the French Revolution. The relationship between Antoine and Marie Anne was of enormous importance for the history of chemistry. Mme Lavoisier trained herself in chemistry to the point of being able to collaborate with her husband. She read and translated English. She studied drawing and engraved the thirteen copperplate illustrations for her husband's famous text, *An Elementary Treatise on Chemistry*. She also assisted Antoine in the laboratory, recording the results of experiments, and was one of the most energetic promoters of the "new chemistry" that resulted from that work.

Lavoisier thus received a great deal of help close to home, and one of the earliest problems on which he labored was also a domestic, or at least a municipal, matter. It was also an affair at whose heart a very real alchemical pulse could, even then, be clearly felt. The query and dispute had to do with water, and it began with a proposal to divert water from regional rivers for the use of the population of Paris. In working out how best to determine the potability of water and how to determine what mineral content local waters possessed, experimenters weighed the solid residues remaining in a container after the water had been completely boiled away. The question arose, however, whether the mineral residue had been actually in the water or whether the water had, in part, been transmuted into earth as the water was evaporated or distilled. Transmutation, in other words, still claimed a place among acceptable possible solutions, and the authority of van Helmont and Boyle, as well as the experimental results of a German physician named Johann Theodor Eller, were drawn on in support of an alchemical explanation.

It was at this point that a very young Lavoisier stepped into public and academic view. His conjecture was that some of the solid matter produced during distillation might actually have come from the glass container in which the water was boiled. Rather than boiling the water, he would allow the water to evaporate slowly by using the instrument that we earlier referred to as a pelican, in which water could continuously evaporate and condense within a sealed vessel. The process is today called refluxing. Lavoisier weighed the vessel with its quantity of water and put the pelican into a sand bath for slow heating. After several weeks, he noted the appearance of solid matter on the sides of the vessel. Weighing the instrument with its contents once again, he found no real change of weight. However, after pouring the water and the solid residue into another container, he found that the pelican was lighter than it had been at first. After weighing the solid matter separately, he discovered that the residue was roughly equal to the weight that the pelican had lost. The solid matter (silica) had come from the glass, he concluded, and was not the result of transmutation.

✳ This well-known experiment not only established Lavoisier's reputation within the French Academy of Science, but it also oriented his thinking toward questioning the definition of elements and, ultimately, toward recognizing that air was made up of different gases that were themselves responsible for different chemical reactions. If I do not describe in any more detail this "revolution in chemistry," it is not because I wish to slight the conceptual or methodological innovations that have justly secured Lavoisier a distinguished place in the history of chemistry. My point is simply that to include alchemy and chemistry as parts of the Scientific Revolution, it is not necessary to wait until Lavoisier made use of quantitative (gravimetric) techniques in the laboratory, acknowledged the conservation of weight (already remarked on, albeit in different terms, in the writings of van Helmont), or explained combustion and calcination by means of oxygen. Nor is it necessary to call on the mechanical philosophy as a way of making chemistry "rational" and only then relevant to the history of science. In fact, transcending modern categories of the "rational" and the "scientific" is important in evaluating what truly belongs to natural knowledge in the early modern world. Cleaving matter from spirit may be a notable achievement from the point of view of contemporary experimental research; but to partition the two in the early modern era, so as to separate wholesome science from feeble metaphysics, is to make a serious mistake.

Separating the supposed rational purity of chemistry from the alleged logical impurities of alchemy as a way to establish the compelling features of a new chemical discipline is also misdirected because chemistry itself did not so much replace alchemy as subsume it. Moreover, even if chemistry is viewed as more elegant from an analytic point of view, we should be careful lest our attraction stagger us into ignoring other ways in which manipulating the substances of nature led to new knowledge. After all, just because one thing is appraised more beautiful than another, that does not mean that the beauty of the thing less desired vanishes altogether (cf. Scarry, 1999).

In a disciplinary sense, chemistry is an extract, a derivative of alchemy. As we have seen, chemistry itself, as practical knowledge and concocted experience (as opposed to a philological, historical, or moral category of debate) (cf. Abbri, 2000), first became suitable to the university not by becoming anything new or unique but by adapting itself to the procedures of medieval alchemy and traditional (scholastic) natural philosophy. For that to happen, what was called chemistry in the late sixteenth century had to shed itself of some very unattractive baggage. Among those to help in this regard, I have pointed especially to the writings of a German physician and school teacher named Andreas Libavius. He was one of the earliest to expound a view of "chemistry" outside the sites and communities where what was called *chemia* remained a private, largely noncommunicative subject held mostly in the hands of Paracelsian adepts. And yet, Libavius was himself an alchemist who argued openly and reasonably (at least within the context of natural philosophy based in Aristotle) for the reality of transmutation. Given such circumstances, any vision that would read back into the history of "chemistry" origins dependent on a complete break with earlier alchemical processes and procedures must be apprehended as an illusion at best.

※ Historians of science have sometimes entered into what gets called the Scientific Revolution with a preformed notion of what should count there as natural knowledge and what should be the best way to get it. What is rational and open, it is assumed, is good and what is emotional and private is not. We know, however, that the world of learning is a messy place and that categories like public and private, reason and passion, often overlap. The same is true for language. Science, some say, means clarity, while pseudo-science depends on obscure terms and enigmatic expressions (cf. Dobbs, 1990). And yet this too is problematic when we take seriously the degree to which expressions about the world are embedded in culture. What is assumed to be clear about language is, in this respect, often relative to what one expects to hear. Libavius, for instance, decried the allegorical and symbolic language of some Paracelsian and alchemical writers and called for clear and didactically useful terminology to replace it. The real language of chemistry, he insisted, had to be based in sophisticated Latin with chemical names composed from Greek. To some not trained within the university, this seemed like another kind of secretive language, the kind reserved for an academic elite (Moran, 1998). On the one hand, then, one can find alchemical writers using symbols and enigmatic references who, once one knows how to interpret what they say, offer clear directions for preparing medicines, cosmetics, alloys, and even the Philosophers' Stone. On the other hand, one finds writers who advocate clarity and openness but whose language is considered obscure by those most skilled in practical procedures and the work of the hands. Sometimes, even those places most lauded as the open and accessible locations of experimental science (laboratories and other specialized workshops) become themselves enigmatic locales when they contain unconventional instruments and exhibit marvelous events far removed from the daily experiences of a witnessing public. What is secretive and what is open depends a great deal on one's own cultural perspective (Long, 2001). Now and then, the neat categories of the Scientific Revolution become, when viewed from the inside out, far less distinct than originally depicted.

Science is human and human beings are a muddle. In his book *The Periodic Table*, Primo Levi referred to chemistry as "a mess compounded of stenches, explosions, and small futile mysteries." There too he distinguished two conflicting philosophical conclusions. The one he called "the praise of purity, which protects from evil like a coat of mail." The other he referred to as "the praise of impurity, which gives rise to changes, in other words, to life." "So," he continued, "take the solution of copper sulfate which is in the shelf of reagents, add a drop of it to your sulfuric acid, and you'll see the reaction begin: the zinc wakes up, it is covered with a white fur of hydrogen bubbles, and there we are, the enchantment has taken place" (Levi, 1984: 34, 60).

The enchantment of the Scientific Revolution, I have argued, has much to do with the presence of impurities of various sorts—the sometimes inharmonious intellectual and social mixture of learned and artisan, of occult, spiritual, and mechanical. This is the concoction that woke things up and produced a cultural reaction. Its description in the preceding chapters has been my own praise of impurity. Through the messy mixture of conflict and diversity, alchemical writers extended the repertoire of imaginable opinion. Theirs was a clamorous "voice," a commotion at the interface between reason and passion, theory and practice, belief and experience. That voice has relevance for the Scientific Revolution because, in the examination of nature, the agitation it caused created emotional "shoving power." It also raised questions and challenged the intellect. Leaving this voice unheard in discussions of the Scientific Revolution limits the potential of varieties of experience to offer intellectual options and to find solutions to practical problems. In this regard, to insist on the superiority of a mechanical and mathematical approach to natural knowledge while trying to bring to life the study of nature in the Renaissance and early modern periods would be quite literally to miss the magic.

✳ One of the most frequently used analogies during the era of the Scientific Revolution is the image of nature as a book. However, while the metaphor of the "book of nature" was commonplace, many in the sixteenth and seventeenth centuries believed that the language of that book had changed. To Paracelsus, the mysterious and powerful relationships between the words of nature's book required imagination, experience, and divine "light," or revelation, to comprehend. For Galileo, the language of the book of nature was not composed of letters but made up of "triangles, circles, and other geometrical figures." Kepler, Descartes, Mersenne, Leibniz, Newton, and members of the Royal Society would have agreed that understanding nature required devising a philosophical language based in mathematical symbols. Others split their allegiance. Bacon, for instance, accepted the necessity of applying mathematics to nature while still trusting that basic truths could be communicated in words. For others, the text of nature came together by means of collecting and organizing her parts. Robert Hooke advocated collecting and organizing the objects of nature within the context of a museum as a means of piecing together the words and phrases of nature's book. The book turned out to be a lexicon. There, within a museum-like collection of objects, an inquirer, he writes, "might peruse, and turn over, and spell, and read the Book of Nature, and observe the *Orthography, Etymology, Syntaxis,* and *Prosodia* of nature's grammar, and by which, as with a *Dictionary,* he might readily . . . find the true Figure, Composition, Derivation and Use of the Characters, Words, Phrases and Sentences of Nature written with indelible, and most exact, and most expressive letters" (Hooke, 1971: 338).

What is often regarded as a simple empirical approach to studying the mixtures of substances is analogous to producing a kind of lexicon. Libavius, Brendel, Boyle, and others were well aware of the utility that came as a result of collecting chemical procedures and knew that processes of separation and combination disclosed the letters out of which the compounds, or words of nature, were formed. Process is action. It makes things happen. Sometimes it separates and considers boundaries. Sometimes it combines and queries about connections. It leads to control, to the artificial construction of useful objects and to claims of power. And all these things—processes, practices, as well as theories—are important to the pursuit of natural knowledge. They are all parts of the diversity of learning that helped to create what we call the Scientific Revolution and that affect the doing of science still.

During the sixteenth and seventeenth centuries, an ongoing process involving manipulation, making, empirical surprises, and polemical interpretations joined artisans and scholars together in the pursuit of natural knowledge. Something indeed was happening during this period and some of the most important representatives of the Scientific Revolution were attracted to and became involved with alchemy and chemistry. Boyle and Newton both maintained alchemical research programs, and the astronomer Tycho Brahe, as Jole Shackelford has shown, combined an observatory with a chemical laboratory at his castle, Uraniborg. (Shackelford, 1993). Nevertheless, to conclude that alchemy should have a place in the Scientific Revolution solely because of the company it kept would be to attend too little to alchemy's distinct cultural influence in the early modern world. It is significant that Newton, Boyle, and others already in the conventional metaphor of the Scientific Revolution turned their attention to alchemical readings and labors; but as important as such relationships are in rehabilitating alchemy as a subject worthy of scientific interest, this cannot be the end of the story. If it were, then alchemy might get cut off at the knees to make it fit into a Procrustean bed. Yet alchemy can depend on its own two feet, and we do not have to rid ourselves of the metaphor of the Scientific Revolution for it to do so. After all, in the search for historical structure the term is helpful. Terms, of course, can have several meanings. If, along with specific discoveries and articulated methodologies, the Scientific Revolution also includes within its horizon ways in which processes and practices can count as objects, in which making leads to learning, and in which the messiness of conflict leads to discernment, then alchemy already has its feet well inside the framework of this vital part of the history of science.

Abbri, Ferdinando. 2000. "Alchemy and Chemistry: Chemical Discourses in the Seventeenth Century." *Early Science and Medicine* 5: 214–226.

Abrahams, Harold. 1971. "Introduction." *Book of Distillation by Hieronymus Brunschwig.* New York: Johnson Reprint.

Agricola, Georgius. 1556; rept. 1950. *De Re Metallica,* trans. Herbert Clark Hoover and Lou Henry Hoover. New York: Dover.

Anderson, Robert. 2000. "The Archeology of Chemistry." In *Instruments and Experimentation,* eds. Frederic L. Holmes and Trevor H. Levere. Cambridge, Mass.: MIT Press, 5–34.

Bacon, Francis. 1620; rept. 1960. *The New Organon,* ed. Fulton Anderson. Indianapolis: Bobbs-Merrill.

Bacon, Roger. 1897–1900; rept. 1964. *The "Opus Majus" of Roger Bacon,* ed. with intro. by John Henry Bridges. Frankfurt/Main: Minerva.

Barchusen, J. C. 1698. *Pyrosophia.* Leiden: Impensis Cornelli Boutestein.

Becher, J. J. 1703. *Physica Subterraneorum,* 8th ed. Leipzig: Apud Joh. Ludov. Gleditschium.

Beguin, Jean. 1669; rept. 1983. *Tyrocinium Chymicum,* trans. Richard Russel. Berkeley Heights: Heptangle.

Benzenhöfer, Udo. 1989. *Johannes' de Rupescissa Liber de Consideratione quintae essentiae omnium rerum deutsch.* Stuttgart: Franz Steiner.

Beretta, Marco. 1997. "Humanism and Chemistry: The Spread of Georgius Agricola's Metallurgical Writings," *Nuncius* 12: 2–47.

Biringuccio, Vannoccio. 1943; rept. 1990. *The Pirotechnia of Vannoccio Biringuccio,* trans. Cyril Stanley Smith and Martha Teach Gnudi. New York: Dover.

Boas, Marie. 1976. *Robert Boyle and Seventeenth-Century Chemistry.* Cambridge, England: Cambridge University Press.

———. 1956. "Acid and Alkali in Seventeenth Century Chemistry," *Archives Internationales d'Histoire des Sciences* 35: 13–28.

Boerhaave, Herman. 1735. *Elements of Chemistry,* trans. Timothy Dallowe. London: For J and J Pemberton et al.

Bohn, Johannes. 1696. *Dissertationes chymico-physicae.* Leipzig: Apud J. Thomam Fritsch.

Bollas, Christopher. 1987. *The Shadow of the Object.* New York: Columbia University Press.

Bono, James. 1995. *The Word of God and the Languages of Man.* Madison: University of Wisconsin Press.

Boyle, Robert. 1690. *The Christian Virtuoso.* In the Savoy, Edw. Jones: For John Taylor.

————. 1664; 1st ed. 1663. *Some Considerations Touching the Usefulnesse of Experimental Natural Philosophy.* Oxford: Hen. Hall for Ric. Davis.

————. 1661; rept. 1911. *The Sceptical Chemist.* London: Everyman's Library.

Bridges, John Henry. 1964. "Introduction." In *The "Opus Majus" of Roger Bacon.* Frankfurt/Main: Minerva.

Brunschwig, Hieronymus. ca. 1530. English trans. of Lawrence Andrew, rept. 1971. *Book of Distillation,* intro. by Harold Abrahams. New York: Johnson Reprint.

Bueno, Mar Rey and María Esther Pérez. 2001. "Los destiladores de su Majestad," *Dynamis* 21: 323–350.

Buntz, Herwig. 1971. "Das 'Buch der heiligen Dreifaltigkeit.' Sein Autor und seine Überlieferung," *Anzeiger für deutsches Altertum* 10: 150–160.

Butters, Suzanne. 1996. *The Triumph of Vulcan.* Florence: Leo S. Olschki.

Clericuzio, Antonio. 1995. "Carneades and the Chemists." In *Robert Boyle Reconsidered,* ed. Michael Hunter. Cambridge, England: Cambridge University Press, pp. 79–90.

————. 1993. "From van Helmont to Boyle: A Study of the Transmission of Helmontian Chemical and Medical Theories in Seventeenth-Century England," *Brit. J. Hist. Sci.* 26: 303–334.

————. 1990. "A Redefinition of Boyle's Chemistry and Corpuscular Philosophy," *Annals of Science* 47: 561–589.

Cook, Harold. 1986. *The Decline of the Old Medical Regime.* Ithaca: Cornell University Press.

Cortese, Isabella. 1561. *I Secreti.* Venice: Appresso Giovanni Bariletto.

Crisciani, Chiara. 1973. "The Conception of Alchemy As Expressed in the *Pretiosa Margarita Novella* of Petrus Bonus of Ferrara," *Ambix* 20: 165–181.

Crosland, Maurice. 1996. "Changes in Chemical Concepts and Language in the Seventeenth Century," *Science in Context* 9: 225–240.

Daston, Lorraine. 1988. "The Factual Sensibility," *Isis* 79: 452–467.

Dear, Peter. 1995. *Discipline and Experience*. Chicago: University of Chicago Press.

Debus, Allen. 2001. *Chemistry and Medical Debate: van Helmont to Boerhaave*. Canton, Mass.: Science History Publications.

———. 1991. *The French Paracelsians*. Cambridge, England: Cambridge University Press.

———. 1977; rept. 2002. *The Chemical Philosophy*. Mineola: Dover.

De Meun, Jean. 1971. *The Romance of the Rose*, trans Charles Dahlberg. Princeton: Princeton University Press.

Descartes, René. 1637. *Discourse on Method*, trans. Donald Cress 1980. Indianapolis: Hackett Publishing.

Dobbs, Betty Jo. 1991. *The Janus Face of Genius: The Role of Alchemy in Newton's Thought*. Cambridge, England: Cambridge University Press.

———. 1990. "From the Secrecy of Alchemy to the Openness of Chemistry." In *Solomon's House Revisited*, ed. Tore Fängsmyr. Canton: Science History Publications, pp. 75–94.

———. 1975. *The Foundations of Newton's Alchemy*. Cambridge, England: Cambridge University Press.

Duchesne, Joseph (Quercetanus). 1605. *The Practice of Chymicall, and Hermeticall Physicke*, trans. Thomas Timme. London: By Thomas Cole.

Eamon, William. 1994. *Science and the Secrets of Nature*. Princeton: Princeton University Press.

———. 1984. "Arcana Disclosed: The Advent of Printing, the Book of Secrets Tradition and the Development of Experimental Science in the Sixteenth Century," *History of Science* 22: 111–150.

Engelsing, Rolf. 1973. *Analphabetentum und Lektüre*. Stuttgart: J. B. Metzlersche Verlagsbuchhandlung.

Forbes, Robert James. 1948. *A Short History of the Art of Distillation*. Leiden: Brill.

Frankl, Victor. 1988. *The Will to Meaning*. New York: Meridian.

Freind, John. 1737. *Chymical Lectures*. London: For Aaron Ward.

Frühsorge, Gotthardt and Gerhard Strasser, eds. 1993. *Johann Joachim Becher (1635–1682)*. Wiesbaden: Otto Harrassowitz.

Gause, Ute. 1991. "Zum Frauenbild im Frühwerk des Paracelsus." In *Parerga Paracelsica*, ed. Joachim Telle. Stuttgart: Franz Steiner, pp. 45–56.

Getz, Faye Marie. 1991. "To Prolong Life and Promote Health: Baconian Alchemy and Pharmacy in the English Learned Tradition." In *Health, Disease and Healing in Medieval Culture*, eds. Sheila Campbell et al. New York: St Martin's Press, pp. 141–151.

Ghiselin, Brewster. 1954. *The Creative Process,* 2nd ed. Berkeley: University of California Press.

Giglioni, Guido. 1997. "The Machines of the Body." In *Marcello Malpighi: Anatomist and Physician,* ed. Domenico Bertolini Meli. Florence: Leo S. Olschki, pp. 149–174.

————. 1995. "Automata Compared: Boyle, Leibniz and the Debate on the Notion of Life and Mind," *Brit. J. Hist. Phil.* 3: 249–278.

Gilly, Carlos. 1998. "'*Theophrastia Sancta*'—Paracelsianism as a Religion in Conflict with Established Churches." In *Paracelsus: The Man and his Reputation,* ed. Ole Peter Grell. Leiden: Brill, pp. 151–185.

Glaser, Christophle. 1663. *Traite de la Chymie.* Paris: Chez l'Autheur.

Goetsch, James Robert, Jr. 1995. *Vico's Axioms.* New Haven: Yale University Press.

Golinski, Jan. 1990. "Chemistry in the Scientific Revolution: Problems of Language and Communication." In *Reappraisals of the Scientific Revolution,* ed. David C. Lindberg and Robert S. Westman. Cambridge, England: Cambridge University Press, pp. 367–396.

Greg, Hugh. 1691. *Curiosities in Chymistry.* London: H. C. for Stafford Anson.

Halleux, Robert. 1979. *Les Textes alchimiques.* Tournhout: Brepols.

Hammond, Mitchell. 1998. "The Religious Roots of Paracelsus's Medical Theory," *Archiv für Reformationsgeschichte* 89: 7–21.

Hannaway, Owen. 1975. *The Chemists and the Word.* Baltimore: The Johns Hopkins Press.

Harkness, Deborah. 1999. *John Dee's Conversations with Angels.* Cambridge, England: Cambridge University Press.

Hirsch, Rudolf. 1950. "The Invention of Printing and the Diffusion of Alchemical and Chemical Knowledge," *Chymia* 3: 115–141.

Hoffmann, Friedrich. 1738. *Medicinae rationalis systematicae tomi quarti.* Frankfurt/Main: Francisci Varrentrapp.

————. 1722. *Oberservationum physico-chymicorum selectiorum libri III.* Halae: In officina Libraria Rengeriana.

————. 1695. *Fundamenta Medicinae,* trans. Lester King, 1971. New York: Science History Publications.

Hooke, Robert. 1971. *Posthumous Works of Robert Hooke,* ed. T. M. Brown. London: Frank Cass and Co.

Hunter, Michael. 2000. *Robert Boyle (1627–91): Scrupulosity and Science.* Woodbridge: Boydell Press.

————, ed. 1994. *Robert Boyle Reconsidered.* Cambridge, England: Cambridge University Press.

Ihde, A. J. 1964. "Alchemy in Reverse: Robert Boyle on the Degradation of Gold," *Chymia* 9: 47–57.

Jacob, Margaret. Forthcoming. *Glimpses of the Cosmopolitan in Early Modern Europe.*

———. 1997. *Scientific Culture and the Making of the Industrial West.* Oxford: Oxford University Press.

James, William. 1896; rept. 1979. *The Will to Believe and Other Essays in Popular Philosophy,* eds. Frederick H. Burkhardt et al. Cambridge, Mass.: Harvard University Press.

———. 1909; rept. 1996. "The Compounding of Consciousness." In *A Pluralistic Universe.* Lincoln: University of Nebraska Press, pp. 181–221.

Kahn, Didier. 2001. "Inceste, assasinat, persecutions et alchimie en France et a Génève (1576–1596)," *Bibliothèque d'humanisme et Renaissance* 63: 227–259.

Kaplan, Barbara. 1993. *Divulging the Useful Truths in Physick.* Baltimore: The Johns Hopkins University Press.

Karpenko, Vladimír. 1992. "The Chemistry and Metallurgy of Transmutation," *Ambix* 39: 47–74.

Kent, Andrew and Owen Hannaway. 1960. "Some New Considerations of Beguin and Libavius," *Annals of Science* 16: 241–250.

King, Lester. 1970. *The Road to Medical Enlightenment.* New York: Elsevier.

———. 1969. "Medicine in 1695: Friedrich Hoffmann's *Fundamenta Medicinae*," *Bull. Hist. Med.* 43: 17–29.

———. 1964. "Stahl and Hoffmann: A Study in Eighteenth-Century Animism," *J. Hist. Med.* 19: 118–130.

Lemery, Nicholas. 1698. *A Course of Chymistry.* London: For Walter Kettilby.

———. 1685. *Modern Curiosities of Art and Nature.* London: For Matthew Gillflower.

Levi, Primo. 1984. *The Periodic Table,* trans. Raymond Rosenthal. New York: Schocken.

Libavius, Andreas. 1613–1615. *Syntagmatis arcanorum et commentationem chymicorum partis tertiae.* Frankfurt/Main: Sumptibus Petri Kopffij.

———. 1606. *Alchymia . . . recognita, emendata, et aucta.* Frankfurt/Main: Johannes Saurius, impensis Petri Kopffij.

———. 1597. *Alchemia.* Frankfurt/Main: Iohannes Saurius, impensis Petri Kopffij.

———. 1595–1599. *Rerum chymicarum . . . liber,* in three parts. Frankfurt/Main: Ioannis Saurius, impensis Petri Kopffij.

———. 1594. *Neoparacelsica.* Frankfurt: Ioannis Saurius, impensis Petri Kopffij.

Linden, Stanton J., ed. 1597; rept. 1992. *The Mirror of Alchemy*. New York: Garland.

Long, Pamela. 2001. *Openness, Secrecy, Authorship*. Baltimore: The Johns Hopkins Press.

Markham, Gervase. 1615; rept. 1986. *The English Housewife*, ed. Michael Best. Kingston: McGill-Queen's University Press.

McGuire, J. E. and P. M. Rattansi. 1966. "Newton and the Pipes of Pan," *Notes and Records of the Royal Society* 21: 108–143.

McKee, Francis. 1998. "The Paracelsian Kitchen." In *Paracelsus: The Man and His Reputation*, ed. Ole Peter Grell. Leiden: Brill, pp. 293–308.

Metzger, Hélène. 1937; rept. 1991. *Chemistry*, trans. Colette V. Michael. West Cornwall: Locust Hill Press.

Meurdrac, Marie. 1666; rept. 1999. *La Chymie charitable et facile, en faveur des Dames*, ed. Jean Jacques. Paris: CNRS Editions.

Moran, Bruce. 1998. "Medicine, Alchemy, and the Control of Language." In *Paracelsus: The Man and his Reputation*, ed. Ole Peter Grell. Leiden: Brill, pp. 135–149.

————. 1996a. "A Survey of Chemical Medicine in the 17th Century," *Pharmacy in History* 38: 121–133.

————. 1996b. "Paracelsus, Religion, and Dissent: The Case of Philip Homagius and Georg Zimmermann," *Ambix* 43: 65–79.

————. 1991. *Chemical Pharmacy Enters the University*. Madison: American Institute of the History of Pharmacy.

Müller, Ingo. 1991. *Iatromechanische Theorie und ärztliche Praxis*. Stuttgart, Franz Steiner.

Multhauf, Robert. 1956. "The Significance of Distillation in Renaissance Medical Chemistry," *Bull. Hist. Med.*, 30, 329–46.

————. 1954. "John of Rupescissa and the Origins of Medical Chemistry," *Isis* 45: 357–367.

Newman, William. 2001. "Experimental Corpuscular Theory in Aristotelian Alchemy." In *Late Medieval and Early Modern Corpuscular Matter Theories*, eds. Christoph Lüthy, John E. Murdoch, and William R. Newman. Leiden: Brill, pp. 291–329.

————. 1998. "The Place of Alchemy in the Current Literature on Experiment." In *Experimental Essays—Versuche zum Experiment*, eds. Michael Heidelberger and Friedrich Steinle. Baden-Baden: Nomos, pp. 9–33.

————. 1995. "The Philosophers' Egg: Theory and Practice in the Alchemy of Roger Bacon," *Micrologus* 3: 75–101.

————. 1994a. "Boyle's Debt to Corpuscular Alchemy." In *Robert Boyle Reconsidered*, ed. Michael Hunter. Cambridge, England: Cambridge University Press, pp. 107–117.

————. 1994b. *Gehennical Fire*. Cambridge, Mass.: Harvard University Press.

————. 1991. *The Summa Perfectionis of the Pseudo Geber*. Leiden: Brill.

Newman, William and Lawrence Principe. 2002. *Alchemy Tried in the Fire*. Chicago: Chicago University Press.

Obrist, Barbara. 1986. "Die Alchimie in der mittelalterlichen Gesellschaft." In *Die Alchemie in der europäischen Kultur-und Wissenschaftsgeschichte*, ed. Christoph Meinel. Wiesbaden: Otto Harrassowitz, pp. 33–59.

Ogrinc, H. L. 1980. "Western Society and Alchemy from 1200 to 1500," *Journal of Medieval History* 6: 103–132.

Pagel, Walter. 1982. *Joan Babtiste van Helmont: Reformer of Science and Medicine*. Cambridge, England: Cambridge University Press.

————. 1958. *Paracelsus*. Basel: S. Karger.

Paracelsus. 1922–1933; rept. 1996. *Sämtliche Werke*, eds. Karl Sudhoff and Wilhelm Mattießen, 14 vols. Hildesheim: Georg Olms Verlag.

————. 1949. *Volumen Medicinae Paramirum*, trans. Kurt Leidecker. Baltimore: The Johns Hopkins Press.

Paré, Ambroise. 1573; rept. 1982. *On Monsters and Marvels*, trans. Janis L. Pallister. Chicago: University of Chicago Press.

Patai, Raphael. 1994. *The Jewish Alchemists*. Princeton: Princeton University Press.

Patterson, T. S. 1937. "Jean Beguin and his Tyrocinium Chymicum," *Annals of Science* 2: 243–298.

Pereira, Michela. 2000. "Heavens on Earth: From the 'Tabula Smaradina' to the Alchemical Fifth Essence," *Early Science and Medicine* 5: 131–144.

————. 1999. "Alchemy and the Use of Vernacular Languages in the Late Middle Ages," *Speculum* 74: 336–356.

————. 1998. "*Mater Medicinarum*: English Physicians and the Alchemical Elixir in the Fifteenth Century." In *Medicine from the Black Death to the French Disease*, eds. Roger French et al. Aldershot: Ashgate.

————. 1995a. "Teorie dell'elixir nell'alchemia medievale," *Micrologus* 3: 103–148.

————. 1995b. "Arnnaldo da Villanova e l'alchimia. Un'indagine preliminare," *Arxiu de Textos Catalans Antics* 14: 95–174.

————. 1989. *The Alchemial Corpus Attributed to Raymond Lull*. London: The Warburg Institute.

Powers, John. 1998. "'Ars sine Arte': Nicholas Lemery and the End of Alchemy in Eighteenth-Century France," *Ambix* 45: 163–189.

Principe, Lawrence. 2001. "Wilhelm Homberg: Chymical Corpuscularianism and Chrysopoeia in the Early Eighteeenth Century." In *Late Medieval and Early Modern Corpuscular Matter,* eds. Christoph Lüthy et al. Leiden: Brill, pp. 535–556.

———. 2000. "The Alchemies of Robert Boyle and Isaac Newton." In *Rethinking the Scientific Revolution,* ed. Margaret Osler. Cambridge, England: Cambridge University Press, pp. 201–220.

———. 1998. *The Aspiring Adept.* Princeton: Princeton University Press.

———. 1994. "Boyle's Alchemical Pursuits." In *Robert Boyle Reconsidered,* ed. Michael Hunter, 91–105.

Principe, Lawrence and William Newman. 2001. "Some Problems with the Historiography of Alchemy." In *Secrets of Nature,* eds. William Newman and Anthony Grafton. Cambridge, Mass.: MIT Press, pp. 385–431.

Pumfrey, Stephen. 1998. "The Spagyric Art; or, the Impossible Work of Separating Pure from Impure Paracelsianism." In *Paracelsus: The Man and His Reputation,* ed. Ole Peter Grell. Leiden: Brill, pp. 21–51.

Reti, Ladislao. 1965. "Parting of Gold and Silver with Nitric Acid in a Page of *Codex Atlanticus* of Leonardo da Vinci," *Isis* 56: 307–319.

———. 1952. "Leonardo da Vinci's Experiments on Combustion," *J. Chem. Ed.* 29: 590–596.

Rothschuh, K. E. 1976. "Studien zu Friedrich Hoffmann (1660–1742)," *Sudhoffs Archiv* 60: 163–193; 235–270.

Rutherford, Donald. 1995. *Leibniz and the Rational Order of Nature.* Cambridge, England: Cambridge University Press.

Ryff, Walther Hermann. 1541. Practicir Büchlin der Leibartznei. Frankfurt/Main: Christian Egenolff.

Sarton, George. 1927–1948. *Introduction to the History of Science,* 3 vols. Baltimore: Williams and Wilkens.

Scarry, Elaine. 1999. *On Beauty and Being Just.* Princeton: Princeton University Press.

Shackelford, Jole. 1993. "Tycho Brahe, Laboratory Design, and the Aim of Science," *Isis* 84: 211–230.

Shapin, Steven. 1994. *A Social History of Truth.* Chicago: University of Chicago Press.

Shapin, Steven and Simon Schaffer. 1985. *Leviathan and the Air Pump.* Princeton: Princeton University Press.

Shaw, Peter. 1730. *Philosophical Principles of Universal Chemistry.* London: For John Osborn and Thomas Longman.

Sherley, Thomas. 1672; rept. 1978. *A Philosophical Essay.* New York: Arno Press.

Smith, Pamela. 2004. *The Body of the Artisan.* Chicago: University of Chicago Press.

———. 1994. *The Business of Alchemy.* Princeton: Princeton University Press.

Tachenius, Otto. 1677. *Hippocrates Chymicus . . . with his Clavis,* trans. J. W. London: Nath. Crouch.

Telle, Joachim. 1992. "Bemerkungen zum 'Rosarium philosophorum.'" In *Rosarium Philosophorum: Ein alchemisches Florilegium des Spätmittelaltrs,* vol. 2, ed. Joachim Telle. Weinheim: VCH, pp. 161–201.

———. 1982. *Pharmazie und der gemeine Mann,* ed. Joachim Telle. Braunschweig: Waisenhaus-Buchdruckerei.

———. 1980. *Sol und Luna.* Hürtgenwald: G. Pressler.

Tosi, Lucia. 2001. "Marie Meurdrac: Paracelsian Chemist and Feminist," *Ambix* 48: 69–82.

Van Helmont, Jean Baptiste. 1667; 4th ed. *Ortus Medicinae,* ed. Francisco Mercurio van Helmont. Lyon: Joan. Ant. Huguetan et Guillielmi Barbier.

Vico, Giambattista. 1994. *The New Science,* trans. Thomas Bergin and Max Fisch. Ithaca: Cornell University Press.

Westfall, Richard. 1980. *Never at Rest.* Cambridge, England: Cambridge University Press.

Willis, Thomas. 1681. *A Medical-Philosophical Discourse of Fermentation,* trans. S. P. London: Printed for T. Dring et al.

Wojcik, Jan. 1997. *Robert Boyle and the Limits of Reason.* Cambridge, England: Cambridge University Press.

Yates, Frances. 1964. *Giordano Bruno and the Hermetic Tradition.* London: Routledge.

———. 1954; rept 1982. "The Art of Ramon Lull." In *Lull and Bruno: Collected Essays,* vol. 1. London: Routledge, pp. 9–77.

Young, John. 1998. *Faith, Medical Authority and Natural Philosophy: Johann Moriaen, Reformed Intelligencer, and the Hartlib Circle.* Aldershot: Ashgate.

Sylvius, Franciscus de le Boë, 116–117, 120
Sympathetic attractions, 90, 109

Tachenius, Otto, 117, 120; *Hippocrates the Chemist,* 135–136
Tartar, Paracelsus on, 78
Telle, Joachim, 25, 27, 55
Thickening by evaporation, 129
Thrithemius of Sponheim, 71
Tosi, Lucia, 65
Transmutation, 10, 25–26, 133, 170; of metals, 19, 21, 22–23, 25, 29, 35, 36, 39–40, 42, 70, 72, 87, 93, 114, 125, 126, 128–129, 146–147, 148, 149–150, 153–155; vs. transubstantiation, 32–33; Paracelsus on, 70–71, 74, 128; of food, 74; Duchesne on, 87; and Libavius, 128, 185; Boyle on, 146–147, 183; Stahl on, 153–155
Trinity, the, 83, 87
Tschirnhaus, Count Ehrenfried Walther von, 148

Ulmannus, 33
Universities. *See* Academics
Urine, 129, 130

Valentine, Basil, 166
Van Helmont, Franciscus Mercurius, 91
Van Helmont, Jean Baptiste, 36, 89–94, 115, 163, 184; on sympathetic attractions, 90–91; on diseases, 91, 93; *The Origin of Medicine,* 91, 93; on water as principle of things, 91, 94–95, 140–141, 142, 148–149; on smoke/gas, 91–92; on spiritual seeds (*semina*), 91–92, 94–95, 96, 141–142, 156, 162; on *blas,* 92, 93; on air, 92–93; on knowledge, 92–93; on the Alkahest, 93, 140; tree experiment of,

93, 140, 148–149; on acids and alkalis, 116, 117, 120; on fermentation, 116, 139; and Boyle, 139–141, 147; influence of, 139–142, 147, 162, 183
Vasari, Georgio, 38
Verrocchio, Andrea del, 37–38
Vesalius, Andreas, 84
Vico, Giambattista, 136
Vincelli, Dom Bernardo, 13
Vital heat, 88
Vitalism, 29, 91–92, 93–94, 98, 155–156, 158, 165–166, 174, 178, 181
Vitriol, 29, 38, 46, 65, 153–154

Water: as element/principle of things, 25–26, 72, 85, 91, 94–95, 117, 121–122, 129, 135, 140–141, 142, 143, 145, 148–149, 152; van Helmont on, 91, 94–95, 140–141, 142, 148–149; Greg on, 94; Glaser on, 119, 121; Willis on, 121, 122; Becher on, 148–149, 152; Stahl on, 152; and Lavoisier, 182–183
Weapon salve, 90–91, 93
Webster, John: *Metallographia,* 141–142
Westfall, Richard, 167
William of Dalby, 31
Willis, Thomas, 121–122
Wine: distillation of, 12, 19; fifth essence of, 19, 21
Wine vinegar, 18
Wojcik, Jan, 138
Women, 53, 60–65; and distillation, 62, 63, 65; during pregnancy, 96–97, 156

Yates, Frances, 68
Young, John, 84

Zodiacal signs, 68–69
Zwinger, Theodore, 85